Molecular Biology of Cancer

F. Macdonald

DNA Laboratory, Regional Genetic Services, Birmingham Heartlands Hospital, Yardley
Green Unit, Birmingham B9 5PX, UK

and

C.H.J. Ford

Faculty of Medicine, Kuwait University, PO Box 24923, 13110 Kuwait

βIOS
SCIENTIFIC
PUBLISHERS

© BIOS Scientific Publishers Limited, 1997

First published 1997

A CIP catalogue record for this book is available from the British Library.

ISBN 1 859962 25 4

BIOS Scientific Publishers Ltd
9 Newtec Place, Magdalen Road, Oxford OX4 1RE, UK
Tel. +44 (0)1865 726286. Fax. +44 (0)1865 246823
World Wide Web home page: http://www.Bookshop.co.uk/BIOS/

DISTRIBUTORS

Australia and New Zealand
DA Information Services
648 Whitehorse Road, Mitcham
Victoria 3132

Singapore and South East Asia
Toppan Company (S)PTE Ltd
38 Liu Fang Road, Jurong
Singapore 2262

India
Viva Books Private Limited
4325/3 Ansari Road, Daryaganj
New Delhi 110 002

USA and Canada
BIOS Scientific Publishers
PO Box 605, Herndon
VA 20172-0605

Typeset by Poole Typesetting, Bournemouth, UK.
Printed by Information Press Ltd, Oxford, UK.

Contents

Abbreviations

ABL	Abelson
AIDS	acquired immunodeficiency syndrome
ALL	acute lymphocytic leukemia
AML	acute myeloid leukemia
AP-1	activator protein 1
AT	ataxia telangiectasia
B-CLL	B-cell chronic lymphocytic leukemia
BCR	breakpoint cluster region
bHLHZ	basic helix–loop–helix–zipper
bp	base pairs
BWS	Beckwith–Wiedemann syndrome
CDK	cyclin-dependent kinase
CDK4I	cyclin-dependent kinase 4 inhibitor
CDKH2	cyclin-dependent hormone inhibitor 2 (also called p16 and CDK4I)
CDKN2	cyclin-dependent kinase 4 inhibitor
CDR	complementarity-determining region
CEA	carcinoembryonic antigen
CHRPE	congenital hypertrophy of the retinal pigment epithelium
CLL	chronic lymphocytic leukemia
CMC	chemical mismatch cleavage
CML	chronic myeloid leukemia
CNTF	ciliary neurotrophic factor
CSF	colony stimulating factor
CTL	cytotoxic T-lymphocytes
DAG	diacylglycerol
DAM	DNA adenosine methylase
DDS	Denys–Drash syndrome
DGGE	denaturing gradient gel electrophoresis
DM	double-minute chromosome
DOP-PCR	degenerate oligonucleotide-primed polymerase chain reaction
EBV	Epstein–Barr virus
EGF	epidermal growth factor
EGFR	epidermal growth factor receptor
ERCP	endoscopic retrograde cholangiopancreatography

ERK	extracellular signal-related kinase
FACS	fluorescence-activated cell sorter
FAP	familial adenomatous polyposis
FGF	fibroblast growth factor
FMTC	familial medullary thryoid carcinoma
G_1	first gap phase (of cell cycle)
G_2	second gap phase (of cell cycle)
GAP	GTPase-activating protein
HAMA	human anti-mouse antibodies
HCC	hepatocellular carcinoma
HNPCC	hereditary nonpolyposis colon cancer
HPR	haptoglobin-related
HPV	human papillomavirus
HSR	homogenous-staining region
HSVtk	herpes simplex virus thymidine kinase
IGF2	insulin-like growth factor 2
IL-1	interleukin 1 (2, 3, 4, etc.)
IP3	inositol 1,4,5-triphosphate
LOH	loss of heterozygosity
LTR	long terminal repeat
M	mitosis (of cell cycle)
MALT	mucosa-associated lymphoid tissue
MAP	mitogen-activated protein
MCL	mantle cell lymphoma
MCR	major cluster region
MCR	mutation cluster region (in the *APC* gene)
MDM2	murine double minute-2
MDS	myelodysplastic syndrome
MEN1A	multiple endocrine neoplasia type 1
MEN2A	multiple endocrine neoplasia type 2 A
MHC	major histocompatibility complex
MRD	minimal residual disease
NF2	neurofibromatosis type 2
NFI	neurofibromatosis
NHL	non-Hodgkin's lymphoma
NSCLC	non-small-cell lung cancer
PI3-K	phosphatidylinositol 3-kinase
PCNA	proliferating cell nuclear antigen
PCR	polymerase chain reaction
PDGF	platelet-derived growth factor
Ph'	Philadelphia chromosome
PHL	Philadelphia gene
PIP2	phosphatidylinositol 4,5-biphosphate
PLC	phospholipase C

PSA	prostate-specific antigen
PTT	protein truncation test
RER	replication error
RER	replication error
RFLP	restriction fragment length polymorphism
RT-PCR	reverse transcription polymerase chain reaction
S	stationary phase (of cell cycle)
SCLC	small-cell lung cancer
SH2	*SRC* homology 2
SH3	*SRC* homology 3
SIN	squamous intraepithelial neoplasia
SSCP	single-strand conformation polymorphism analysis
TCC	transitional cell carcinoma
TCR	T-cell receptor
TGFα/β	transforming growth factor α or β
TNM	tumor, nodes, metastasis staging system
TPC	translocated in prostate cancer
VHL	von Hippel–Lindau disease
VNTR	variable number of tandem repeats
WHO	World Health Organisation

Preface

Since the publication of the first edition of this book, *Oncogenes and Tumor Suppressor Genes*, there have been significant advances not only in the study of these two groups of genes but also in our knowledge of the other genes involved in the development of the malignant phenotype. This is reflected in the altered title of the second edition of the book. The aim of this second edition is still to provide nonspecialists in the field, including medical students, postgraduates and medical practitioners, with a readable text which both summarizes the scientific aspects of these genes and shows how they can be used to influence patient management. As in the first edition, it is not the intention to cover the biology of the various genes in great depth, as there are many excellent texts available which do this. Instead the aim is to provide the reader with sufficient background information to understand the subsequent chapters on the clinical use of the genes. The introductory chapters have been expanded to cover the genes involved in cell cycle control as well as the mismatch repair genes. Reflecting the increasing information available on the use of molecular genetics in cancer diagnosis and treatment, the major cancers are covered in separate chapters. Potential therapeutic applications, including gene therapy, are covered in the penultimate chapter and finally the chapter on techniques used to study these genes has been updated to include newer technologies, particularly with reference to mutation detection.

Our thanks go to Dr Alan Cockayne, Dr Alan Casson and Dr Andrew Read for reading parts of the manuscript and for helpful comments. We'd also like to thank those colleagues who provided us with many of the figures and to BIOS Scientific Publishers for their help. Finally we have to thank our families, whose lives have been disrupted over recent months, for their forbearance which allowed us to complete this manuscript.

F. Macdonald
C.H.J. Ford

Chapter 1

General principles

1.1 Introduction

It has been realized for many years that cancer has a genetic component and at the level of the cell it can be said to be a genetic disease. In 1914, Boveri suggested that an aberration in the genome might be responsible for the origins of cancer. This was subsequently supported by the evidence that cancer, or the risk of cancer, could be inherited; that mutagens could cause tumors in both animals and humans; and that tumors are monoclonal in origin, that is, the cells of a tumor all show the genetic characteristics of the original transformed cell. It is only in recent years that the involvement of specific genes has been demonstrated at the molecular level.

Cancer cells contain many alterations which accumulate as tumors develop. Over the last 20 years, considerable information has been gathered on regulation of cell growth and proliferation leading to the identification of the proto-oncogenes and the tumor suppressor genes. The proto-oncogenes encode proteins which are components of the cell signaling pathways. Mutations in these genes act dominantly and lead to a gain in function accelerating cell division. In contrast the tumor suppressor genes inhibit cell proliferation by arresting progression through the cell cycle and block differentiation. They are recessive at the level of the cell although they show a dominant mode of inheritance. Mutations leading to increased genomic instability suggest defects in mismatch and excision repair pathways. We also now know that the genes involved in regulating the cell cycle are also abnormally expressed in tumors. Finally, there is a group of genes involved in DNA repair which, when mutated, predispose the patient to developing cancer. This failure of DNA repair is seen in xeroderma pigmentosum, ataxia telangiectasia, Fanconi's anemia and Bloom's syndrome [1]. In addition, many other genes encoding proteins, such as proteinases or other enzymes capable of disrupting tissues, and vascular permeability factors have been shown to be involved in carcinogenesis. Epigenetic events such as alterations in the degree of methylation of DNA have also been detected in tumors [2]. Genomic imprinting, a process in which expression of two alleles is dependent on the parent of origin, has also been described in cancers [3]. Any combination of these changes may be found in an individual tumor. The overall progression to malignancy is therefore a complex event.

1.2   What is cancer?

In normal cell growth there is a finely controlled balance between growth-promoting and growth-restraining signals such that proliferation occurs only when required. The balance is tilted when increased cell numbers are required, for example during wound healing and during normal tissue turnover. Differentiation of cells during this process occurs in an ordered manner and proliferation ceases when no longer required. In tumor cells this process is disrupted, continued cell proliferation occurs and loss of differentiation may be found. In addition the normal process of programed cell death may no longer operate.

Cancers arise from a single cell which has undergone mutation. Mutations in genes such as those described in the next three chapters give the cell increased growth advantages compared to others and allows them to escape normal controls on proliferation. The initial mutation will cause cells to divide to produce a genetically homogenous clone. In turn additional mutations occur which further enhance the cells' growth potential. These mutations give rise to subclones within the tumor each with differing properties so that most tumors are heterogenous. This multistep process is described further in Section 1.6.

Tumors can be divided into two main groups, benign or malignant. Benign tumors are rarely life threatening, grow within a well-defined capsule which limits their size and maintain the characteristics of the cell of origin and are thus usually well differentiated. Malignant tumors invade surrounding tissues and spread to different areas of the body to generate further growths or metastases. It is this process which is often the most life threatening. Different clones within a tumor will have differing abilities to metastasize, a property which is genetically determined. The process of metastasis is likely to involve several different steps and only a few clones within a tumor will have all of these properties. Some of the genes involved in the process of metastasis are described in subsequent chapters but additional factors such as proteinases, the cell adhesion molecules E-cadherin [4] and the integrin family have been implicated in the invasive process.

Tumor cells show a number of features which differentiate them from normal cells: (1) They are no longer as dependent on growth factors as normal cells either because they are capable of secreting their own growth factors to stimulate their own proliferation, a process termed autocrine stimulation, or because growth factor receptors on the surface are altered in such a way that binding of growth factors is no longer necessary to stimulate proliferation; (2) normal cells require contact with the surface in the extracellular environment to be able to grow whereas tumor cells are anchorage independent; (3) normal cells respond to the presence of other cells, and in culture will form a monolayer due to contact inhibition whereas tumor cells lack this and often grow over or under each other; (4) tumor cells are less adhesive than normal cells; (5) normal cells stop proliferating once they reach a certain density but tumor cells continue to proliferate.

1.3 The cell cycle

Central to the complex process of proliferation is the control of the processes involved in driving the cell through the cell cycle. The cell cycle involves a series of events which result in DNA duplication and cell division. In normal cells this process is carefully controlled but in tumor cells, mutations in the genes associated with the cell cycle result in progression of cells with damaged DNA through the cycle. The phases of the cell cycle have been known for around 40 years but it is only in the last 3 or 4 years that the genes which are involved in control of the cycling process have been identified. The cell cycle is divided into four distinct phases. The first gap phase (G_1), DNA replication (S) and the second gap phase (G_2) together make up interphase, the period from the end of one nuclear division to the start of the next. This is followed by mitosis or M (*Figure 1.1*).The length of the cell cycle varies considerably from one species and tissue to another but is typically around 16–24 h. In G_1, cells are preparing to synthesize DNA and biosynthesis of both RNA and proteins occurs. The length of this phase is the most variable. During S phase, DNA is replicated and histones are synthesized. At the end of S phase the DNA content of the cell has doubled and the chromosomes have replicated. In G_2, cells are preparing for cell division, the replicated DNA complexes with proteins and biosynthesis continues. The nucleus and cytoplasm finally divide during mito-

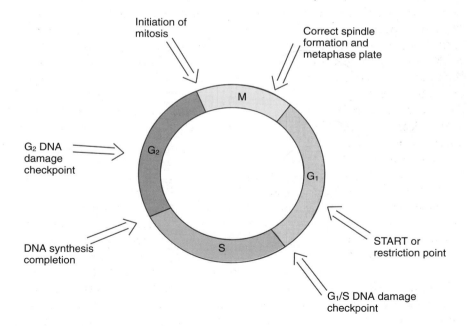

Figure 1.1: The cell cycle showing checkpoints at which DNA is monitored before the next stage of the cycle is entered.

sis and two daughter cells are produced which can then begin interphase of a new cell cycle if conditions are suitable for further growth. In the absence of mitogens or when nutrients are depleted, cells can also enter a resting phase termed G_0. Similarly, cells which are terminally differentiated, for example neurons, are generally arrested in G_0. Transition from one stage to the next is regulated at a number of checkpoints which prevent premature entry into the next phase of the cycle (*Figure 1.1*).

Mammalian cells initially respond to external stimuli such as growth factors to enter into G_1. Once cells have become committed to replication, that is towards the end of G_1, they become refractory to growth factor-induced signals and to growth factor inhibitors such as TGFβ and then rely on the action of the cell cycle proteins to regulate progression. This progression from one stage to the next is carefully controlled by the sequential formation, activation and subsequent degradation or modification of a series of cyclins and their partners, the cyclin-dependent kinases (CDKs; see Chapter 4) [5]. In addition it has been shown more recently that a further group of proteins, the cyclin-dependent kinase inhibitors (CDKIs) are important for signal transduction and coordination of each stage of the cell cycle.

There are a number of recognized checkpoints at which the integrity of the DNA and the formation of the spindle and spindle pole are checked. Defects in the DNA itself can lead to chromosomal rearrangements and transmission of damaged DNA. Defects in the spindle lead to mitotic nondisjunction, a process which results in the gain or loss of whole chromosomes. Defects in the spindle pole can lead to alterations in ploidy [6]. The first checkpoint in mammalian cells is known as START or restriction (R) point and occurs late in G_1. It is at this stage that cells must commit themselves to a further round of DNA replication and at this point any damage to DNA can be monitored and if necessary the cycle halted until it is repaired. Loss of this checkpoint can lead to genomic instability and survival of damaged cells. Further checkpoints occur during S phase and at the G_2 to M transition to monitor the completion of DNA synthesis and the formation of a functional spindle. This last checkpoint prevents chromosome segregation if the chromosomes are not intact. The degradation of various cyclins occurs at each checkpoint and it is this mechanism together with interaction of the CDKIs which allows the cell to enter the next phase.

Initially it was believed that the basic cell cycle machinary would not be involved in tumorigenesis but it is now clear that this is wrong. Checkpoints are deregulated in tumors and mutations occur in the genes involved in cell cycle components, leading to genomic instability and cancer. The mode of action of cyclins, CDKs and CDKIs are discussed in Chapter 4 and abnormalities associated with specific cancers are detailed in subsequent chapters.

1.4 Apoptosis

Apoptosis or programed cell death, first described in 1972, is a series of genetically controlled events which result in the removal of unwanted cells, for example during morphogenesis of the embryo, without disruption of tissues [7]. As with differentiation or proliferation, apoptosis is an important method of cellular control and any disruption of this process, not surprisingly, leads to abnormal growth. It plays an important role in tumorigenesis as well as in many other disease processes such as degenerative and autoimmune diseases. In recent years the field has expanded rapidly and many extensive reviews have been written [8–10], therefore only an overview of the process will be presented here.

Apoptosis differs from necrosis or accidental cell death in a number of important ways. Firstly, it is an active process as opposed to an unplanned process induced by cell injury. Secondly, apoptotic cells are recognized by phagocytes and removed before they disintegrate. As a consequence, there is no surrounding tissue damage or induction of inflammatory responses. In contrast, in necrosis, cells become leaky, release macromolecules and rapidly disintegrate, thereby inducing inflammation.

Apoptotic cells show a very characteristic morphology. They show condensation of nuclear heterochromatin, cell shrinkage and loss of positional organization of organelles in the cytoplasm. Electrophoresis of DNA extracted from apoptotic cells shows a characteristic laddering pattern of oligonucleosomal fragments resulting from internucleosomal chromatin cleavage by endogenous endonucleases. This differs from the smear of degraded DNA usually seen in necrotic cells.

In contrast to the well established morphological features, the genetic events leading to apoptosis are only now beginning to be recognized [9]. A number of genes and proteins, some acting within the cells themselves and others acting extrinsically, have now been identified which either promote or inhibit apoptosis (*Table 1.1*) [11].

Both proto-oncogenes and tumor suppressor genes have been shown to be

Table 1.1: Proteins controlling apoptosis

	Promoting	Inhibiting
Intrinsic	p53 MYC Interleukin-1β converting enzyme BAX/BCLX$_S$	BCL2/BCLX$_L$ A20
Extrinsic	TNFα TGFβ	Many. For example: erythropoietin PDGF/IGF1 sex hormones

Derived from ref. 11.

involved in the control of apoptosis. The *MYC* gene has a dual role in cells. As described in Chapter 2, MYC, complexed with its protein partner MAX, has a role in driving cell proliferation. In addition, it plays a role in committing cells to enter the apoptotic pathway again complexed to MAX. The choice of which pathway to follow appears to depend on the cell's microenvironment, particularly the availability of growth factors [12]. In the absence of suitable growth factors the result is apoptosis. Overexpression or deregulation of *MYC* in tumors therefore has important implications for the control of apoptosis. The tumor suppressor gene *p53* is also involved in the induction of apoptosis and acts by driving cells with damaged DNA along this pathway rather than allowing them to continue through the cell cycle. The importance of this gene has been demonstrated in 'knockout' mice in which both copies of *p53* are absent. A high rate of tumors is seen in these animals and in addition they are highly resistant to the induction of apoptosis [12]. However these 'knockout' mice are developmentally normal and can still undergo apoptosis in response to other forms of genotoxic damage, suggesting that although *p53* plays a role in mediating the response to cell damage it is not an obligate component of the apoptotic pathway.

Both *MYC* and *p53* play a role in the induction of apoptosis. Another proto-oncogene, the *BCL2* gene, acts to block apoptosis specifically [9,12]. Its deregulation will therefore remove suppression of programed cell death and can promote tumor formation. The *BCL2* gene was localized to chromosome 18, at the site of one of the breakpoints of a reciprocal translocation, also involving chromosome 14, in B-cell follicular lymphomas. The *BCL2* gene is widely expressed during embryonic development but in adults is confined to the immature and stem cell populations. It is now known to belong to a large family of related proteins, some of which (e.g. BCL2, $BCLX_L$ and MCL1) suppress apoptosis whereas others (e.g. BAX and $BCLX_S$) act to antagonize the anti-apoptotic properties of the others.

Inhibition of cell death, either by suppression of those genes which induce cell death or by activation of those genes which cause cell survival, therefore contributes to the development of tumors. There is considerable interest in this area not just to gain a better understanding of tumorigenesis but also because it may in the future reveal novel targets for the therapy of cancers.

1.5   Chromosomes and cancer

Normal human cells contain 46 chromosomes. Changes in this number as well as structural chromosomal abnormalities are common in the majority of tumors. The first consistent chromosome abnormality to be recognized was the Philadelphia chromosome seen in chronic myeloid leukemia (see Sections 2.2.2 and 10.5). Since then many other changes have been found, including loss or gain of whole chromosomes or parts of chromosomes and chromosomal translocations, examples of which are given in *Tables 1.2* and *1.3* [13].

These changes are nonrandom events. Some may be primary events occurring early in the development of the tumor and are likely to be an important event in its development. Others are secondary events and may have a role in the subsequent behavior of the tumor. In addition, many other random changes in the chromosome complement are also found due to the instability of the tumor cell.

Table 1.2: Examples of primary chromosomal aberrations in hematological malignancies

Tumor	Abnormality	Tumor	Abnormality
AML	t(1;7)(p11;p11)	ALL	t(1;11)(p32;q23)
	Trisomy 4		t(1;19)(q23;p13)
	Monosomy 5		t(4;11)(q21;q23)
	t(6;11)(q27;q23)		t(8;14)(q24;q32)
	Monsomy 7		t(8;22)(q24;q11)
	Trisomy 8		t(2;8)(p12;q24)
	t(9;11)(p21;q23)		t(9;22)(q34;q11)
	t(10;11)(p14;q13)		t(10;14)(q24;q11)
	Trisomy 11		t(11;19)(q23;p13)
	del/t(12p)		del(12)(p11p13)
	t(15;17)(q22;q11)		t(11;14)(p13;q11)
	inv(16)(p13q22)		t(1;19)(q23;p13)
	del(16)(q22;q24)		Trisomy 21
	i(17q)	CML	t(9;22)(q34;q11)
	del(20)(q11q13)	CLL	Trisomy 12
	Trisomy 21	Lymphoma	t(8;14)(q24;q32)
	Trisomy 22		t(8;22)(q24;q11)
			t(2;8)(p12;q24)

Derived from ref.13. (This list shows examples of some of the rearrangements found in hematological malignancies.) AML, acute myeloid leukemia; ALL, acute lymphocytic leukemia; CML, chronic myeloid leukemia; t, translocation; inv, inversion; del, deletion; i, isochromosome.

Table 1.3: Examples of primary chromosomal aberrations in solid tumors

Tumor	Abnormality
Lipoma	t(3;12)(q27–28;q14–15)
Ewing's sarcoma	t(11;22)(q24;q12)
Renal carcinoma	t or del(3)(p11–21)
Wilms' tumor	t or del(11)(p13)
Bladder carcinoma	Changes of chromosome 1, i(5p)
Breast cancer	Changes of chromosome 1, t or del(16q)
Ovarian cancer	Changes of chromosome 1
Germ cell tumors of testis	i(12p)
Meningioma	Monosomy 22, del(22q)
Neuroblastoma	del(1)(p13–32)
Retinoblastoma	del(13)(q14)
Malignant melanoma	t or del(6q)/i(6p)
	t or del(1)(p12–22)
Uterine carcinoma	Changes of chromosome 1

Derived from ref.13. t, translocation; inv, inversion; del, deletion; i, isochromosome.

Loss of chromosomal material can often result in the deletion of a tumor suppressor gene (see *Figure 3.5*). Duplication of a region can lead to over-expression of an oncogene (see *Figure 2.4*). Studies of chromosome transloca-tions led to the identification of a number of oncogenes, for example, the 8;14 translocation found in Burkitt's lymphoma and shown to be associated with the *MYC* oncogene on chromosome 8 (see Section 2.2.2). Now that these chromo-some abnormalities have been well studied, some, such as the presence of the 9;22 translocation in CML, can be used diagnostically (see Section 10.5).

Constitutional chromosome abnormalities, present in all the cells of the body, have helped in the identification of a number of tumor suppressor genes. In general these changes are found in only one or a few patients but have been instrumental in the subsequent isolation of the gene. The classic examples, dis-cussed more fully in Chapter 3, are the deletions found in retinoblastoma patients leading to the identification of the *RB1* gene; the deletions associated with Wilms' tumor leading to the identification of *WT1*; and the deletion found on the long arm of chromosome 5 which led to the identification of the *APC* gene in familial adenomatous polyposis.

1.6   Inherited vs. sporadic cancer

Cancers can be classified into four main groups on the basis of the genetic defect:

(1) the majority of cancers are sporadic and are caused by environmental fac-tors such as chemicals and radiation. Mutations in these tumors are found only in the cancer tissue itself.

(2) Some cancers, without a recognizable genetic basis, show clustering in families and may represent an underlying susceptibility to environmental carcinogens. Care has to be taken when looking at this group as it is possi-ble to have apparent clustering of cancers in families due to the shared envi-ronment of family members rather than because of a genetic defect. This second group can be recognized by an earlier age of onset, multiple cancers in individuals and by segregation of the disease through the family in a Mendelian manner.

(3) A small proportion of cancers have a clearly defined genetic cause. This means that screening at risk family members is immediately possible, there-by preventing unnecessary morbidity and mortality. In addition, the genes involved have a wider importance as the genes which cause the inherited form of the disease are often the same as those which are implicated in the sporadic form of the disease. Their study can therefore help in the under-standing of the more common forms of the cancer. The classic example of this is the *APC* gene which is responsible for the inherited condition famil-ial adenomatous polyposis (FAP) and which is also the earliest gene to be

mutated in the development of sporadic colorectal cancer (see Section 3.3.2). The main exception to this principle is familial breast cancer which is caused by defects in the *BRCA1* and *BRCA2* genes (see Section 3.3.5). Defects in these genes have not so far been observed in sporadic breast cancer. In cancers with a true genetic basis, the causative mutation is found in all the cells of the body.

(4) Individuals with some conditions, generally termed chromosome breakage syndromes because of the increased chromosome fragility seen in cultured cells, for example those from xeroderma pigmentosum and ataxia telangiectasia, have an increased risk of cancers although the incidence of cancers is not close to the levels seen in patients in group 3.

1.7 The multistage nature of cancer development

In 1949, Berenblum and Shublik concluded that "the recognition that carcinogenesis is at least a two-stage process should invariably be borne in mind" [14]. Armitage and Doll took this observation a step further; in 1954, they published

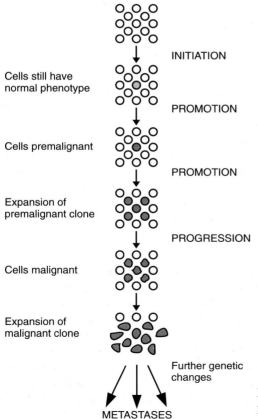

INITIATION

Cells still have normal phenotype

PROMOTION

Cells premalignant

PROMOTION

Expansion of premalignant clone

PROGRESSION

Cells malignant

Expansion of malignant clone

Further genetic changes

METASTASES

Figure 1.2: Multistage progression to malignancy.

age/incidence curves for 17 common types of cancer. From their figures they concluded that carcinogenesis was at least a six or seven stage process [15]. Although each of these steps cannot usually be clearly defined in an individual tumor, it is clear today that there is without doubt a multistage progression to malignancy (*Figure 1.2*).

Tumors tend to acquire more aggressive characteristics as they develop, and in 1957 Foulds pointed out that tumor progression occurred in a stepwise fashion, each step determined by the activation, mutation or loss of specific genes [16]. Over the next two decades biochemical and cytogenetic studies demonstrated the sequential appearance of subpopulations of cells within a tumor, attributable, in part at least, to changes in the genes themselves.

The evidence suggests that, in the majority of cases, cancers arise from a single cell which has acquired some heritable form of growth advantage [17]. This initiation step is believed to be caused frequently by some form of genotoxic agent such as radiation or a chemical carcinogen. The cells at this stage, although altered at the DNA level, are phenotypically normal. Further mutational events involving genes responsible for control of cell growth lead to the emergence of clones with additional properties associated with tumor cell progression. Finally, additional changes allow the outgrowth of clones with metastatic potential. Each of these successive events is likely to make the cell more unstable so that the risk of subsequent changes increases. Animal models of carcinogenesis, primarily based on models of skin cancer development in mice, have enabled these steps to be divided into initiation events, promotion, malignant transformation and metastasis [17] (*Figure 1.2*).

Although it is clear that multiple changes are necessary for tumor development, it is not clear whether the order in which the changes occur is critical. Evidence suggests, however, that it is the accumulation of events that is important rather than the order in which they occur (see Section 3.6.1 for the classic example in colorectal cancer).

References

1. Sancar, A. (1994) *Science*, **266**, 1954.
2. Counts, J.I. and Goodman, J.I. (1995) *Cell*, **83**, 13.
3. Feinberg, A.P. (1993) *Nature Genetics*, **4**,110.
4. Berx, G., Staes, K., van Hengel, J. *et al.* (1995) *Genomics*, **26**, 281.
5. Pines, J (1995) *Adv. Cancer Res.*, **66**, 181.
6. Hartwell, L.H. and Kastan, M.B. (1994) *Science*, **266**, 1821.
7. Kerr, J.F.R., Wyllie, A.H. and Currie, A.R. (1972) *Br. J. Cancer*, **26**, 239.
8. Gregory, C.D. (1993) in *Regulation of the Proliferation of Neoplastic Cells* (L. Pusztai, C.E. Lewis and E. Yap, eds). Oxford University Press, Oxford, p.342.
9. Wyllie, A.H. (1995) *Curr. Opin. Genet. Devel.*, **5**, 97.
10. Clarke, E.R., Sphyris, N. and Harrison, D.J. (1996) *Mol. Med. Today*, **2**, 189.
11. Saville, J. (1994) *Eur. J. Clin. Invest.*, **24**, 715.
12. Harrington, E.A., Fanidi, A. and Evan, G.I. (1994) *Curr. Opin. Genet. Dev.*, **4**, 120.

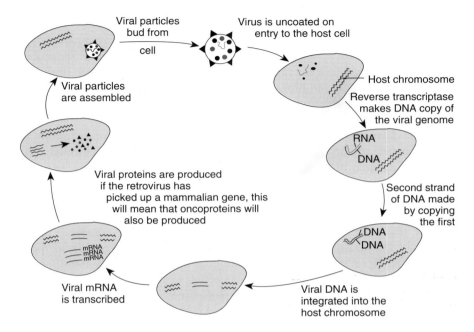

Figure 2.2: The retrovirus life cycle.

because the gene is brought under the control of a viral promoter, leading to aberrant expression. With reference to their potential function in tumor development, the original cellular genes were termed proto-oncogenes. This mechanism of activation is implicated in a wide range of animal tumors but it has never been shown convincingly to be associated with human cancers.

Retroviruses can also activate proto-oncogenes more directly by a process known as insertional mutagenesis. In this process, the insertion of a DNA copy of the retrovirus into the cellular genome close to a proto-oncogene causes abnormal activation of that gene by stimulation of gene expression via the promoter action of the LTRs. This has been demonstrated for the *INT1* gene which is activated in breast cancer in mice infected with the mouse mammary tumor virus. Most of the proto-oncogenes identified in this way are identical to those already found via transforming retroviruses, although a few additional genes such as *EVI1* have also been found.

As few retroviruses had been shown to be the cause of human cancers, it was still not clear how these genes associated with the retroviruses might relate to the pathogenesis of cancer in man. It was only when it was shown that human tumors contained activated oncogenes homologous to those found in the retroviruses, but with no viral intermediary, that this whole area of research expanded rapidly.

2.2 Cellular oncogenes

Whilst research into the retroviral oncogenes continued, more direct methods of identifying oncogenic sequences in the human genome were examined. A DNA transfection assay was used as a method of identifying those sequences in tumor cells which were responsible for uncontrolled cell proliferation. DNA was extracted from human tumors and sheared into fragments. These were then transfected into a mouse-derived cell line called NIH-3T3 so that random fragments were incorporated into its genome. As a result, some cells were transformed and could be identified by their loss of contact inhibition which caused cells to pile up *in vitro* (*Figure 2.3*). These transformed cells were also capable of producing tumors when injected into athymic (nude) mice. The genome of the transformed cells was analyzed and shown to contain an oncogene which in many cases was similar to one which had been identified in the retroviruses [2] (*Table 2.3*). This transfection assay did not give a positive result with all tumors. Only about 20% of tumors contained oncogenes which could be identified in this way and about one-quarter of these belonged to the *RAS* gene family. Additional oncogenes were identified by alternative strategies.

It had long been known that some tumors carry a consistent chromosome translocation. In the case of chronic myeloid leukemia (CML), this is a reciprocal translocation between chromosomes 9 and 22 (see Section 2.2.2). In Burkitt's lymphoma, there is a reciprocal translocation between chromosome 8 and, in the majority of cases, chromosome 14, although either chromosome 2 or chromosome 22 is occasionally involved. In both examples the breakpoints were shown to coincide with the location of oncogenes already identified from retroviral studies; *ABL* on chromosome 9 and *MYC* on chromosome 8. Other translocations in tumors identified the sites of further oncogenes (*Table 2.3*). In a comprehensive study of 5345 tumors [3], the distribution of cancer-specific breakpoints was compared with the site of 26 cellular oncogenes; of these, 19 were located at cancer-associated chromosome breakpoints.

Two other chromosome abnormalities observed in tumors have also identified the location of oncogenes (*Table 2.3*). Both the development of homogeneous-staining regions (HSRs) and formation of double-minute chromosomes (DMs, *Figure 2.4*) were associated with oncogene amplification. Some oncogenes identified in this way, for example, *MYC*, had previously been detected by other techniques, but other genes such as *NMYC*, associated with neuroblastomas, and *LMYC*, found in small cell lung carcinomas, were first revealed in this way [4].

Currently about 60–100 different cellular proto-oncogenes have been identified, which when activated by one of the mechanisms described below give that gene its oncogenic activity (see Appendix A for the location of some of these).

The importance of these genes to the cell is clear as there is sequence conservation from organisms such as yeast, through the invertebrates and vertebrates to

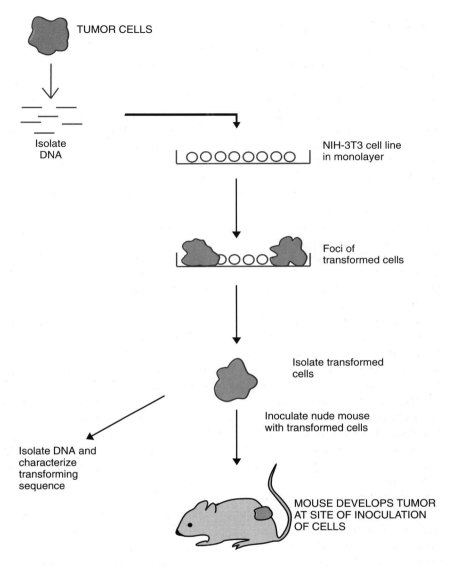

Figure 2.3: DNA transfection assay.

Table 2.3:   Methods of oncogene identification in human tumors

Method of identification	Oncogene
Amplification	*ERBB2, LMYC, NMYC*
Chromosomal translocation	*ABL, BCL1, BCL2, MYC*
Homology to retroviruses	*HRAS, KRAS, SRC*
Insertional mutagenesis	*EVI1, INT1*
Transfection assay	*MAS, MET, MYC, RAS, TRK*

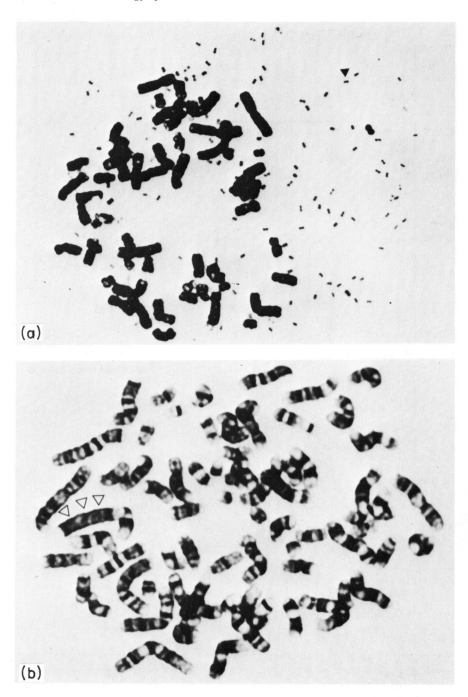

Figure 2.4:   Examples of (a) DMs and (b) HSRs (arrowed). Figure courtesy of Dr J. Waters, Regional Cytogenetics Laboratory, Birmingham Heartlands Hospital, UK.

man. In the normal cell, the expression of these proto-oncogenes is tightly controlled and they are transcribed at the appropriate stages of growth and development of cells. However, alterations in these genes or their control sequences lead to inappropriate expression. What therefore are the functions of these genes and how are they converted into genes capable of contributing to malignant progression?

2.2.1 Function of the proto-oncogenes

Many studies have supported the prediction that the proto-oncogenes would be involved in the basic essential functions of the cell related to control of cell proliferation and differentiation. Cells are stimulated by external signals, growth factors, which bind to cell surface receptors, for example the transmembrane protein kinases, thereby activating their function as signal transducers. In turn this stimulates intracellular signaling pathways, eventually leading to alterations in gene expression. The proto-oncogenes function at each of these steps and are found therefore at all levels of the cell (*Figure 2.5*). Mutations in any of the genes result in their abnormal activation promoting cell growth in the absence of external stimuli and leads to malignant transformation.

Growth factors.   Growth factors constitute the products of the first group of proto-oncogenes and exist as polypeptides, oligopeptides or steroid hormones (*Table 2.4*) [5]. They bind to their own specific receptors or, in a few cases, cross

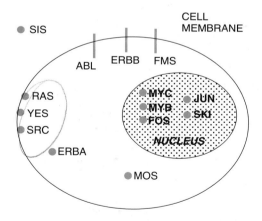

Figure 2.5: Cellular locations of oncogene products.

Table 2.4:   Role of proto-oncogenes in the cell: growth factors

Platelet-derived growth factor (PDGF)
Epidermal growth factor (EGF)
Fibroblast growth factor 1–7 (FGF1–7)
Insulin-like growth factor 1 and 2 (IGF1 and 2)
Transforming growth factor α and β (TGFα and β)

react with a number of receptors, to stimulate or inhibit cell growth acting via a number of pathways culminating in alterations in gene expression. Those which stimulate growth do so either by advancing the cell from the G_0 phase of the cell cycle into G_1 (e.g. epidermal growth factor (EGF), platelet-derived growth factor (PDGF), fibroblast growth factor (FGF)) or by aiding progression through G_1 (e.g. insulin-like growth factor). In contrast, transforming growth factor β (TGFβ) acts as an anti-mitogen by reversibly inhibiting cells at G_1, a process involving activation of a cyclin-dependent kinase inhibitor and in-activation of cyclin/cyclin-dependent kinase activity (see Chapter 4).

Two models of growth factor action have been recognized. In the autocrine model [6], binding of a growth factor to its receptor leads either to increased secretion of more growth factor from the same cell or up-regulation of the receptor promoting autostimulation of tumor growth. This model has been proposed, for example, for the way bombesin secretion acts in small cell lung cancer. In the paracrine model, a growth factor released from one cell (e.g. stromal tissue cells), binds to a receptor on a neighboring cell (e.g. adjacent tumor cells). This in turn stimulates release of a paracrine factor to increase more growth factor production from the first cell. There is therefore reciprocal stimulation of growth factor production.

Growth factor receptors. A second group of proto-oncogenes encode either the growth factor receptors themselves or their functional homologs. The growth factor receptors (*Table 2.5*) link the information from the extracellular environment (the growth factors) to a number of different intracellular signaling pathways. The most important of the growth factor receptors with respect to malignant transformation are the transmembrane receptor tyrosine kinases although other receptors such as the hemopoiesis growth factor receptors and steroid receptors also play a role [5].

Table 2.5:   Role of proto-oncogenes in the cell: growth factor receptors

ERBB1	ERBB2 HER2/NEU	FMS	ROS
TRK	IGF-1R	RET	EEK
BEK	MET	SEA	KIT
FGFR1–4	FIG	EPH	ELK

The receptor tyrosine kinases possess an extracellular, ligand binding domain, a transmembrane domain and one or two intracellular protein tyrosine kinase domains (*Figure 2.6*). Binding of the growth factor to the extracellular domain induces the receptor to dimerize resulting in an increase in the kinase activity of the receptor. This leads to autophosphorylation of tyrosine residues of the receptor itself as well as of other intracellular proteins. The activated receptor then acts as a center for the assembly of a 'signal particle' on the inner surface of the membrane which transmits the signal to the nucleus (*Figure 2.7*).

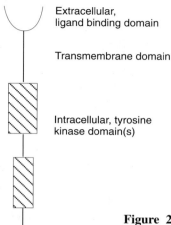

Extracellular,
ligand binding domain

Transmembrane domain

Intracellular, tyrosine
kinase domain(s)

Figure 2.6: Structure of the transmembrane tyrosine kinase receptors showing the three domains.

Signal transducers.   The proteins which are initially recruited in response to tyrosine phosphorylation are Src homology 2 (SH2) and SH3 domain containing proteins. These domains composed of 50–100 amino acids were described in the *SRC* family of proto-oncogenes but have also been found in other unrelated proteins, all of which are capable of interacting with activated protein tyrosine kinases and are involved in cell signaling. These proteins fall into two groups; those with enzymatic activity such as the cytoplasmic tyrosine kinases (e.g. *SRC* and *ABL, Table 2.6*) and phospholipase C (PLC) or a group of adaptor proteins [e.g. phosphatidylinositol 3-kinase (PI3-K) and Grb2]. The SH2 and SH3 domains recognize specific amino acid sequences in their target proteins. SH2 domains bind preferentially to tyrosine phosphorylated residues whereas SH3 domains recognize amino acid sequences containing proline and hydrophobic residues and may promote binding to membranes or the cytoskeleton. There are three main ways in which receptor tyrosine kinases activate the SH2/SH3 proteins (*Figure 2.8*) leading to three downstream signaling pathways [7].

PI3-K is an enzyme composed of two subunits, p85 and p110, thought to have an important role in the relay of signals from the growth factor receptors. The binding of the p85 subunit via its SH2 domains, to the tyrosine phosphorylated residues of the receptor (e.g. PDGF) results in a conformational change in p85 which is transmitted to the p110 catalytic subunit. This results in increased enzyme activity although the exact mechanism by which it sends a signal is not fully understood.

The SH2 domains of PLC bind to tyrosine phosphorylated EGF receptor, resulting in phosphorylation of PLC itself and hence its activation. It is then capable of hydrolyzing membrane-associated phosphatidylinositol 4,5-biphosphate (PIP2) to two second messengers, diacylglycerol (DAG) and inositol 1,4,5-triphosphate (IP3). These in turn can stimulate the activation of protein

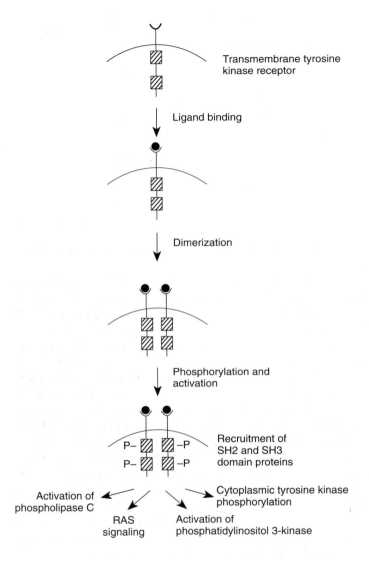

Figure 2.7: Activation of the transmembrane tyrosine kinase receptors leading to four downstream signaling pathways.

kinase C, an important intermediate in signaling pathways, and also cause release of stored intracellular calcium.

The SH2 domain of the adaptor molecule, Grb2, binds to tyrosine phosphorylated proteins such as the EGF receptor recruiting the RAS guanine nucleotide-releasing factor called Sos (which stands for *son of sevenless*, the *Drosophila* equivalent of the gene) to the cell membrane in close proximity to RAS. Sos is bound to Grp2 via its SH3 domain. Interaction of inactive RAS

Table 2.6:   Role of proto-oncogenes in the cell: signal transducers

Membrane associated/guanine nucleotide binding proteins	RAS
	GSP
	GIP
Membrane associated/cytoplasmic protein tyrosine kinases	ABL
	SRC
	FES
	FGR
	SYN
	LCK
Cytoplasmic protein serine–threonine kinases	RAF
	MOS
	COT

with Sos leads to GDP/GTP exchange, thereby activating the next step of the signaling pathway.

RAS is one of the best studied of the signal transducers [8]. The main members of the *RAS* gene family, *HRAS*, *KRAS* and *NRAS* each code for a protein, p21, which can bind GTP and exhibits GTPase activity. When bound to GDP, RAS p21 is inactive but active when bound to GTP. RAS p21 proteins in their active form are capable of activating the mitogen activated protein (MAP) kinase cascade and thereby promoting DNA synthesis. The first step of this process is the recruitment of the serine/threonine kinase, RAF, to the plasma membrane where it is activated. In turn it phosphorylates and activates its substrate MEK (MAP/ERK kinase) which can phosphorylate and activate a further kinase, ERK (extracellular signal-related kinase) leading to phosphorylation of nuclear proteins and DNA synthesis (*Figure 2.9*). In addition, RAS p21 in its active state can also activate PI3-K initiating the alternative pathway involving RAC and RHO. These GTP binding proteins each regulate a signal transduction pathway linking growth factor receptors to changes in the actin cytoskeleton. Active RAS p21 is converted back to the inactive GDP bound form by the GTPase activating proteins (GAPs). At least two GAPs exist in humans. The first to be identified was a 120 kDa protein called p120-GAP, the second was the product of the neurofibromatosis type 1 (*NF1*) gene, neurofibromin, which shows homology to a region of about 300 amino acids in p120-GAP. *NF1* is believed to be a tumor suppressor gene (see Chapter 3) and therefore shows an example of the interaction between proto-oncogenes and tumor suppressor genes.

A further group of signal transducers are the cytoplasmic tyrosine kinases such as ABL and SRC. SRC kinase possesses multiple domains including a site of amino-terminal myristylation, SH2 and SH3 domains, a kinase domain and a carboxyl-terminal regulatory domain. SRC is maintained in its inactive state by phosphorylation of a tyrosine residue at position 527 in the carboxyl-terminal domain and binding of this tyrosine to the SH2 domain. In addition, the SH3 domain has been shown to be necessary to repress the intrinsic kinase

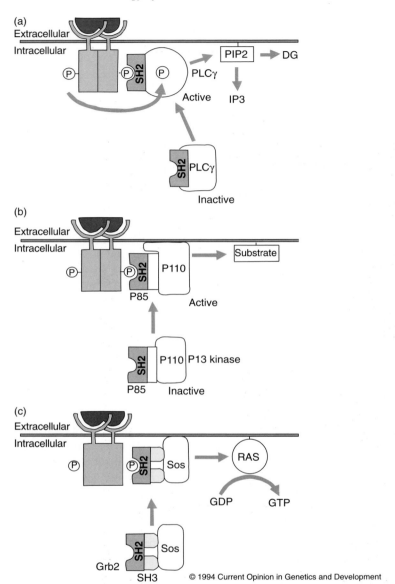

Figure 2.8: Mechanism of activation of SH2/SH3 signaling molecules. (a) Activation of phospholipase C. The SH2 domains of PLCγ-bind to the tyrosine phosphorylated EGF receptor. facilitating PLCγ phosphorylation and hence its activation. (b) Interaction of the tyrosine phosphorylated kinase domain of PDGFR and the SH2 domain of the p85 sub-unit of PI3-K promotes a conformation change in p85 which is transmitted to the p110 catalytic subunit of PI3-K and enhances its activity. (c) The two SH3 domains of Grb2 bind to the guanine nucleotide releasing factor, Sos. This complex binds to the tyrosine phosphorylated EGF receptor thereby translocating Sos to the membrane. Interaction of Sos leads to GDP/GTP exchange on RAS and GTP bound RAS then activates a kinase cascade. Reproduced from ref. 7 with permission from Current Biology Ltd.

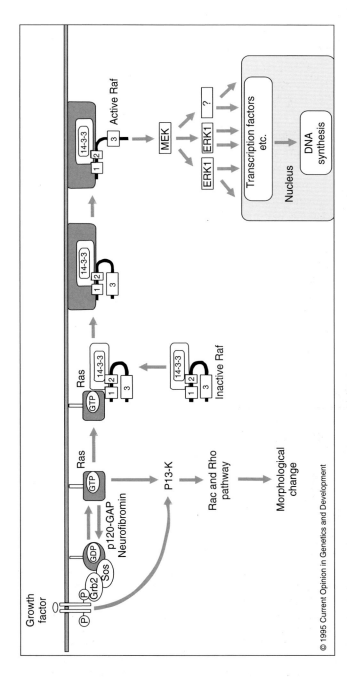

Figure 2.9: Signal transduction from receptor to nucleus via RAS p21. Reproduced from ref. 8 with permission from Current Biology Ltd.

Table 2.7:   Role of proto-oncogenes in
the cell – oncogenes with nuclear effects

MYC
FOS
JUN
MYB
SKI
EVI1
REL

activity of SRC, possibly binding to residues in the kinase domain. Activation of SRC kinases occurs following their association with receptor tyrosine kinases such as PDGFR. The SH3 domain of SRC has been shown to bind the p85 regulatory subunit of PI3-K, a mechanism by which SRC may activate this molecule in addition to the mechanism described above via the receptor tyrosine kinases.

The *ABL* proto-oncogene has many features in common with *SRC*. The amino terminal of the protein contains both SH2 and SH3 domains and also a region with tyrosine kinase activity. The major function of the carboxyl-terminal domain is in subcellular localization and it has been shown that ABL binds to F actin. In addition, this region contains a DNA binding domain. If the functions of these domains are put together it has been suggested that the tyrosine kinase and SH2 domains are involved in signaling pathways occuring on actin filaments and/or on DNA [9]. In addition, a protein homologous to GAP, 3BP-1, has been identified which interacts with the SH3 domain suggesting that this region might link ABL to signaling pathways involving GTP binding proteins.

Nuclear proto-oncogenes and transcription factors.   The final group of proto-oncogenes are those involved in the control of gene expression by their action on DNA itself. This is the final site of action for messages sent from the growth factors and is the level at which control of growth and proliferation ultimately operates. Several of the proto-oncogene proteins have been shown to bind to DNA and presumably control the transcription of genes [e.g. *MYC*, *FOS* and *JUN* (*Table 2.7*)].

The *MYC* gene family (including *c-MYC*, *NMYC* and *LMYC*) are believed to be essential for cell proliferation and prevention of differentiation in response to mitogenic stimuli. The MYC protein, p62, possesses two domains, an amino-terminal domain with transactivational activity and a carboxyl-terminal DNA binding and dimerization domain with a 'leucine zipper' motif (bHLHZ, standing for basic helix–loop–helix–zipper) similar to other transcription factors. It is now known that the MYC protein dimerizes *in vitro* with a second protein, MAX [10]. MYC–MAX heterodimers function as sequence specific transcriptional regulators and this dimerization is necessary for subsequent DNA binding to occur *in vivo*. Binding of MYC–MAX heterodimers occurs at a specific DNA sequence motif, CACGTG, which has been found in the promoter of several genes shown to be regulated by *MYC*. Over-expression of MAX leads to suppression of MYC-MAX dimer activity, suggesting that

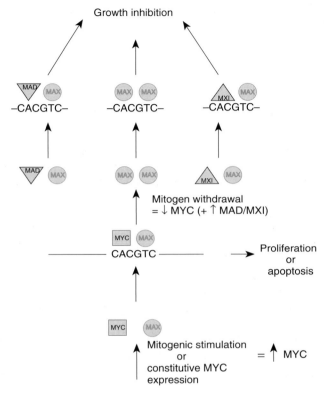

Figure 2.10: Mechanism for the role of MYC, MAX, MAD and MXI in the control of cell growth

MAX is a suppressor of *MYC* activity as well as an activator.

Two further proteins, MAD and MXI-1, have recently been identified which also form heterodimers with MAX and which bind to and compete for the same hexanucleotide binding motif as MYC–MAX. Changes in the expression of the genes encoding these proteins, in response to the presence or absence of growth stimuli, leads to alterations in the levels of each of the proteins and consequently alters the composition of the dimers. MAX is a stable protein which is constitutively expressed whereas the other three are rapidly degraded in cells. Stimulation of cells leads to the increase in expression of *MYC* and hence to the production of MYC–MAX dimers. Withdrawal of mitogens or the removal of differentiation stimuli leads to the suppression of *MYC* expression and to an increase in the alternative dimers, MAX–MAD and MAX–MXI-1. The assumption is that these dimers are responsible for growth inhibition (*Figure 2.10*).

In addition, *MYC* is recognized as an inducer of programed cell death or apoptosis [11]. This process also requires dimerization with MAX and is likely to involve a transcriptional mechanism with modulation of specific genes. *MYC* therefore plays a role in two tightly coupled pathways, one resulting in proliferation and the other in apoptosis.

A further class of transcription factors thought to serve as the nuclear target for mitogenic signals are the proteins which belong to the activator protein 1 (AP-1) family. Two members of this group are the *JUN* and *FOS* gene families, whose expression in cells has been shown to be elevated following exposure to growth factors such as PDGF and EGF. JUN and FOS have been shown to form heterodimers, the two protein products being held together by a 'leucine zipper'. These dimers regulate transcription by binding to DNA at AP-1 sites, which are regulatory sites found in a variety of genes associated with proliferation and differentiation [12].

As described above, the various groups of proto-oncogenes perform major functions in the cell. It might therefore be expected that any disruption in their expression would lead to disruption in the control of that cell, leading to its death, if the proto-oncogene function is lost, or to transformation if the proto-oncogene functions excessively or inappropriately. The various mechanisms by which the proto-oncogenes are activated to produce a gene with oncogenic potential are described in the next section.

2.2.2   Mechanisms of oncogene activation

Table 2.8 shows the three main ways by which proto-oncogenes are activated. The first mechanism is production of an abnormal product which can occur in a number of ways. Point mutations have been described in several oncogenes but have been studied most extensively in genes of the *RAS* family. Amino acid substitutions have been detected, particularly at positions 12, 13 and 61, in a variety of tumors including breast, lung and colon. Activation of the protein to an oncogenic form is also possible at positions 56, 63, 116 and 119, but these changes have not been seen in human tumors. These substitutions all alter the structure of the normal protein resulting in abnormal activity. Oncogenic forms of *RAS* produce a protein with decreased GTPase activity which prevents deactivation of the RAS–GTP complex, leading to prolonged stimulation of signals from the growth factor receptors [8].

A point mutation in the transmembrane domain of the *NEU* proto-oncogene, resulting in a valine to glutamic acid substitution, causes receptor dimerization and kinase activation in the absence of ligand, a mechanism which may be involved in receptor tyrosine kinase activity [13]. However, these kinases can also be activated in a number of other ways.

Table 2.8:   Mechanisms of oncogene activation

Method of activation	Oncogene
Structural alteration	*ABL, HRAS, KRAS, SRC*
Amplification	*ERBB2, MYB, MYC*
Loss of control mechanism	
Insertional mutagenesis	*INT1, INT2*
Translocation	*MYC*

Oncogenes can also be activated by chromosomal translocation resulting in the production of a fusion protein [14]. One of the first examples of this was in chronic myeloid leukemia where chromosome 9 is translocated to chromosome 22. This places the *ABL* oncogene on chromosome 9 next to the breakpoint cluster region (BCR) of the Philadelphia gene on chromosome 22 (*Figures 2.11* and *10.2*). A fusion gene is formed which in turn produces an abnormal fusion protein. The fusion protein shows constitutively activated tyrosine kinase activity mediated by the interaction of two domains of BCR with ABL. Domain 1 causes oligomerization of BCR–ABL promoting the tyrosine phosphorylation of domain II. This phosphorylation allows the fusion protein to interact with an SH2–SH3 adaptor molecule coupling BCR–ABL to downstream targets in the *RAS* signaling pathway [13]. Subsequently, many other examples of translocations resulting in the creation of fusion genes have been identified and in many cases, although by no means solely, these have been shown to involve transcription factors leading to disruption of transcriptional control [14]. One such example is seen in rhabdomyosarcoma in which a translocation between the long arms of chromosomes 2 and 13 is found. This translocation involves the *PAX3* gene, a developmental control gene encoding a transcription factor, and *FKHR*, a gene which again belongs to a transcription factor family known as the 'forkhead' family. A fusion gene is produced which is made up of the DNA binding domain of *PAX3*, a truncated part of the *FKHR* DNA binding domain and the C-terminal part of *FKHR* [14]. The fusion gene product can bind to PAX3 target sequences and aberrantly regulate transcription (see Section 11.9.2).

Several of the receptor tyrosine kinases encoded by proto-oncogenes (e.g. by *MET* and *RET*) are activated following rearrangements which juxtapose novel sequences derived from unrelated loci with the kinase domain of the receptor, deleting the extracellular ligand binding domain and resulting in the cytoplasmic localization of the proteins [13]. This process results in dimerization of the truncated receptor in the absence of the ligand, a process which requires the interaction of the novel amino acid sequence which is fused with the intracellular domain.

Another cause of oncogenic mutation of proto-oncogenes can be deletion of part of the protein. If the SH2 domain of the SRC protein kinase is deleted, the tyrosine residue at position 527, which is required for maintenance of the inactive state, is removed. SRC can also be activated by replacement of the tyrosine residue with another amino acid such as phenylalanine. Similarly, deletion of the SH3 domain, required to regulate negatively the kinase activity of SRC, also results in a protein with increased tyrosine kinase activity [15].

The second mechanism of activation is over-production of the normal protein by amplification of the proto-oncogene [4]. For example, amplification of the *MYC* gene is found in some breast and colorectal cancers. Amplication of the *NMYC* gene is consistently associated with late-stage neuroblastomas. In breast cancer, amplification of the *ERBB2* gene is a common finding although

(a)

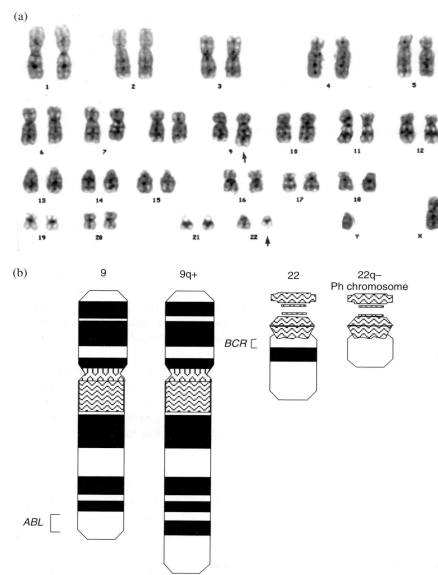

(b)

Figure 2.11   (a) Karyotype from a case of CML showing the Ph' chromosome (arrowed on bottom row). Figure courtesy of Dr S. Hill, Regional Cytogenetics Laboratory, Birmingham Heartlands Hospital, UK. (b) Ideogram of the t(9;22).

there appears to be considerably higher protein levels than would be predicted from gene copy number alone, suggesting that other factors in addition to amplification are also functioning. The *RAS* oncogene, which is frequently activated by point mutation, has also been found to be amplified in certain human tumors.

The third mechanism of activation occurs through loss of appropriate control mechanisms. This has been described in an earlier section for the retroviral oncogenes, where insertional mutagenesis causes transcriptional activation. Cellular oncogenes also show loss of normal transcription control through chromosomal translocation typified by the 8;14 translocation seen in Burkitt's lymphoma [16] (*Figures 2.12* and *10.1*). Following the juxtaposition of the *MYC* oncogene on chromosome 8 to one of the immunoglobulin loci on chromosome 14, 2 or 22, there is constitutive expression of the transposed *MYC* gene. Depending on the position of the breakpoint within the *MYC* gene, overexpression of *MYC* from the translocated allele can occur via at least three mechanisms; activation of cryptic promoters in intron 1 driven by adjacent immunoglobulin elements; transcription driven by the immunoglobulin enhancer in cases where the entire *MYC* transcription unit is translocated; or mutations in exon 1 of *MYC* which can result in the loss of the block to transcription elongation. As with amplification of the proto-oncogenes, the effect of these mechanisms is the over-production of the gene product, resulting in a continuous stimulus for cell proliferation.

There is no single consistent mechanism of activation of any one oncogene. Thus *RAS* is primarily activated by point mutation but amplification is also found, *MYC* is amplified in many tumors but abnormal expression is also associated with deregulation following chromosomal translocation. Whatever the mechanism for the activation of these genes, the end result is to produce a protein which can cause abnormal growth.

Although the main area of research is in how these genes relate to the development of tumors, abnormal products of the proto-oncogenes are also associated with nonmalignant conditions such as arthritis and proliferative vasculitis and are also involved in the process of wound healing where proliferation or differentiation are either required or found to be disordered.

2.3   Oncogenes in human tumors

In most groups of tumors so far examined for the presence of an oncogene, such a gene has been found. A few of the most common tumors and associated oncogenes are shown in *Table 2.9* but the list is far from complete. The value of these oncogenes as diagnostic or prognostic markers is discussed fully from Chapter 5 onwards. Recently it has been shown that the *RET* oncogene is the causative gene in an inherited cancer syndrome, multiple endocrine neoplasia type 2A (MEN2A)[17]. Mutations in proto-oncogenes are very rarely the cause of familial cancers as they are likely to be lethal. In many tumors such as Burkitt's lymphoma or CML, the presence of the oncogene is likely to have an important, although not necessarily primary, role in the development of the tumor. In the majority of other tumors it is not clear exactly what role the presence of the oncogene has to play in the development of the cancer. Some oncogenes are likely to be detected not because they have been specifically activated, but because control of normal cell growth has been disrupted, lead-

(a)

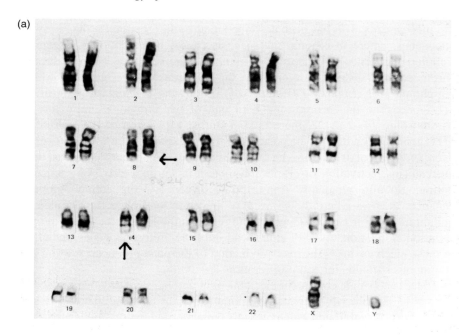

(b)

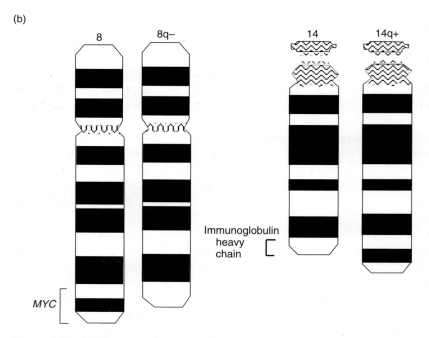

Figure 2.12: (a) Karyotype from a Burkitt's lymphoma containing a t(8;14).
(b) Ideogram of the translocation. Figure courtesy of the Regional Cytogenetics
Laboratory, Birmingham Heartlands Hospital, UK.

Table 2.9: Oncogenes in human tumors

Tumor	Associated oncogene
Bladder	*HRAS, KRAS*
Brain	*ERBB1, SIS*
Breast	*ERBB2, HRAS, MYC*
Cervical	*MYC*
Colorectal	*HRAS, KRAS, MYB, MYC*
Gastric	*ERBB1, HST, MYB, MYC, NRAS, YES*
Lung	*ERBB1, HRAS, KRAS, MYC, LMYC, NMYC*
Melanoma	*HRAS*
Neuroblastoma	*NMYC*
Ovarian	*ERBB2, KRAS*
Pancreas	*KRAS, MYC*
Prostate	*MYC*
Testicular	*MYC*
Leukemia	*ABL, MYC, BCR, BCL1, BCL2*

ing to their subsequent activation. For example, in tumors formed following inoculation of mice with a fibroblast cell line previously transfected with human *MYC* or mutated *HRAS*, elevated levels of the transfected gene could be detected, but there was also elevated expression of the endogenous *ABL* and *FOS* genes. This result suggests that one gene might be able to destabilize the cell sufficiently to provide an environment in which further changes in other oncogenes are likely [18].

Here we need to return to the concept of cancer as a phenotype which arises following a number of sequential alterations in the cell. We stated earlier that a single mutational event was not enough for a tumor to develop, but we have described examples of oncogenes which act in a dominant manner and which in some cases have been detected only by their ability to transform the NIH-3T3 cell line. This experiment needs to be seen in perspective. The NIH-3T3 cell line is not normal but has undergone a number of mutational events during its establishment in culture. Insertion of a single oncogene into a totally normal cell line will not cause it to transform without considerable manipulation of its expression. Much evidence now suggests that there is collaboration between different oncogenes and that this is necessary to produce the fully transformed phenotype.

In model systems it was initially demonstrated that whereas a single oncogene was insufficient for transformation, collaboration between genes such as *RAS* and *MYC* or *RAS* and *p53* (see Chapter 3) could result in full transformation. Some of the observations have been demonstrated more directly in transgenic mice. In this model, a fragment of DNA containing an oncogene linked to a suitable promoter can be injected into a fertilized mouse egg. Often this recombinant fragment of DNA is integrated into the mouse genome and may be expressed in many tissues or only a few depending on the tissue specificity of the promoter. When this was tried with either *MYC* or *RAS* alone, few of the progeny produced tumors, although the introduced gene was generally expressed. Only the occasional cell which had undergone further changes became transformed. Not all the progeny from a cross between two transgenic mice, one carrying *MYC* and the other *RAS*, developed tumors, although an increase in the number of tumors was seen compared with either parent [19].

This interaction between the oncogenes is only one of the processes involved in malignant development. We now know that a further group of genes, which act in a growth-regulatory rather than a growth-promoting fashion, are also involved with each other and with the oncogenes. These genes are the subject of the next chapter.

References

1.   Stehelin, D., Varmus, H.V., Bishop, J.M. and Vogt, P.K. (1976) *Nature*, **260**, 170.
2.   Shih, C., Padhy, L.C., Murray, M. and Weinberg, R.A. (1981) *Nature*, **290**, 261.
3.   Heim, S. and Mitelman, F. (1987) *Hum. Genet.*, **75**, 70.
4.   Schwab, M. and Amler, L.C. (1990) *Genes, Chrom. Cancer*, **1**, 181.
5.   Aaronson, S.A. (1991) *Science*, **254**, 1146.
6.   Sporn, M.B. and Todaro, G.J.(1980) *N. Engl. J. Med.*, **303**, 878.
7.   Schlessinger, J. (1994) *Curr. Opin. Genet. Dev.*, **4**, 25.
8.   McCormick, F. (1995) *Curr. Opin. Genet. Dev.*, **5**, 51.
9.   Wang, J. (1993) *Curr. Opin. Genet. Dev.*, **3**, 35.
10.   Amati, B. and Land, H. (1994) *Curr. Opin. Genet. Dev.*, **4**, 102.
11.   Harrington, E.A., Bennett M.R., Fanidi, A and Evan, G.I. (1994) *EMBO J.*, **13**, 3286.
12.   Angel, P. and Karin, M. (1991) *Biochim. Biophys. Acta*, **1072**, 129.
13.   Rodrigues, G.A. and Park, M. (1994) *Curr. Opin. Genet. Dev.*, **4**, 15.
14.   Rabbitts, T (1994) *Nature*, **372**, 143.
15.   Seidel-Dugan, C., Meyer, B.E., Thomas, S.M. (1992) *Mol. Cell Biol.*, **12**, 1835.
16.   Spencer C.A. and Groudine M. (1991) *Adv. Cancer Res.*, **56**, 1.
17.   Goodfellow, P.J. (1994) *Curr. Opin. Genet. Dev.*, **4**, 446.
18.   Wyllie, A.H., Rose, K.A., Morris, R.G., Steel, C.M., Foster, E. and Spandidos, D.A. (1987) *Br. J. Cancer*, **56**, 251.
19.   Sinn, E. (1987) *Cell*, **49**, 465.

Further reading

Mendelsohn J., Howley, P.M., Israel, M.A. and Liotta, L.A. (1995) *The Molecular Basis of Cancer*. W.B. Saunders Co., Philadelphia.
Vile, R.G. (1992) *Oncogenes – Introduction to the Molecular Genetics of Cancer*. John Wiley & Sons, London.

Chapter 3

Tumor suppressor genes

3.1 Introduction

A second group of genes which plays an important role in tumorigenesis is the tumor suppressor genes. These are defined as genes involved in the control of abnormal cell proliferation and whose loss or inactivation is associated with the development of malignancy. The literature abounds with different names for these genes which may cause some confusion including; ortho-genes, emero-genes, flatogenes and onco-suppressor genes. Most commonly, they are described as tumor suppressor genes or anti-oncogenes. As the genes do not always act directly on an oncogene, this latter term is a misnomer. The term tumor suppressor gene also has its limitations; as the control of growth is likely to involve a number of genetic mechanisms, it is not clear that a single gene can abrogate all of them. However, until the debate on the correct terminology reaches its conclusion this term will be used here.

3.2 Evidence for the existence of tumor suppressor genes

Tumor suppressor genes are more difficult to identify than oncogenes. Introduction of an oncogene into an untransformed cell culture and identification of the resulting transformed colonies is relatively straightforward. It is not as easy to identify untransformed revertants on a background of transformed cells. Two different techniques have been used to establish the existence of tumor suppressor genes in man.

The best studied examples of tumor suppressor genes are found in the fruit fly, *Drosophila melanogaster*. Over 25 different genes which cause tumors by homozygous recessive mutations have been identified. Some of these have been cloned and one, *lethal (2) giant larvae*, shows suppression of tumor formation when re-introduced into the germ-line.

Evidence from two types of study supports the existence of tumor suppressor genes in man. These are: (1) the suppression of malignancy in somatic cell hybrids, and (2) a consistent loss of chromosomal regions, initially seen in hereditary cancers and subsequently also shown in sporadic cancers.

3.2.1   Suppression of malignancy by cell fusion

The earliest evidence for tumor suppressor genes pre-dates the discovery of oncogenes by over 20 years. Harris and colleagues showed that when malignant cells were fused with normal diploid cells, the resulting hybrids were non-tumorigenic as determined by their inability to grow in immunocompromised hosts [1] (*Figure 3.1*). This suppression of malignancy was dependent on retention of a specific chromosome. As the hybrids were unstable, there was random loss of chromosomes and when a particular chromosome was lost, malignant clones were once again capable of growth *in vivo*. The assumption was that the chromosome which was lost contained the tumor suppressor gene(s). Fusion of two malignant cells also gave rise to nontumorigenic hybrids suggesting complementation between lesions. Detailed cytogenetic analysis of hybrids has identified specific chromosomes involved in the suppression of the malignant phenotype. For example, the presence of chromosome 11 from the normal partner is necessary to maintain the suppression of the malignant phenotype when HeLa cervical carcinoma cells are fused with normal fibroblasts. This result was confirmed by introducing a single fibroblast chromosome 11 into HeLa cells by microcell fusion; the malignant phenotype of the cervical cancer cells was suppressed [2].

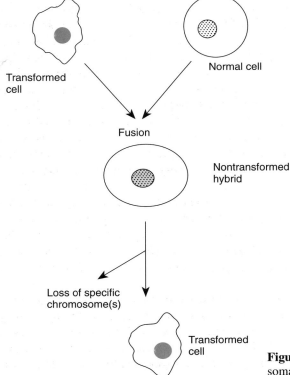

Transformed cell

Normal cell

Fusion

Nontransformed hybrid

Loss of specific chromosome(s)

Transformed cell

Figure 3.1:   Production of somatic cell hybrids.

Although tumorigenesis is suppressed in hybrids between HeLa cells and normal fibroblasts, the hybrids still behave as transformed cells *in vitro*. This suggests that transformation and tumorigenicity are under separate genetic control. Other characteristics of tumor cells, such as immortalization and metastatic ability, can also be suppressed in hybrids, suggesting that a number of different genes are involved in the tumorigenic phenotype [2].

Reversion of the malignant phenotype can also be demonstrated in the presence of an activated oncogene. Nontumorigenic hybrids have been produced following fusion of cell lines carrying either activated *NRAS* or *HRAS* genes with normal fibroblasts. These phenotypically normal hybrids, and also tumorigenic revertants, still express the product of the respective oncogene [2].

3.2.2 Tumor suppressor genes in hereditary cancers: the retinoblastoma model

A second piece of evidence for the existence of tumor suppressor genes came from studies of hereditary cancers. These are cancers where there is a clear pattern of inheritance, usually autosomal dominant, with a tendency for earlier age of onset than for sporadic tumors.

The prototype for studies on hereditary cancers is retinoblastoma, a childhood cancer which occurs in two forms and affects the retina (*Figure 3.2*). Forty percent of cases are hereditary, transmitted as an autosomal dominant trait with 90% penetrance [3]. Of these cases around 10–15% are transmitted from an affected parent, the remainder arising as *de novo* germ-line mutations, more usually on the paternal germ-line than the maternal. Tumors frequently arise in both eyes, the average number of tumor foci being three to five. The remaining 60% of cases are sporadic and characteristically tumors are seen in only one eye. In 1971, from studies of age/incidence curves, Knudson [4] postulated that the disease arose from two sequential mutational events. In the hereditary form of the disease, one mutation is inherited in the germ-line and is phenotypically harmless, confirming the recessive nature of the mutation at the cellular level. A second 'hit' occurring in a retinal cell causes the tumor to develop (*Figure 3.3*). As there are a large number of retinoblasts in the eye (over 10^7), which are all at risk because they already carry one mutation, a second 'hit' will occur frequently enough to cause a high proportion of tumors in at least one eye and often in both. In the sporadic form of the disease, both mutations occur in the somatic tissue (*Figure 3.3*). The probability of two mutations occurring in the same cell is low, therefore the disease is unilateral [5].

This 'two-hit' hypothesis has subsequently been confirmed by identification of mutations or deletions of the gene and more recently by analysis of the cloned retinoblastoma gene itself. Before such an analysis was possible, the retinoblastoma locus had to be mapped to a specific chromosome. A number of patients were identified with a cytogenetically visible deletion in the

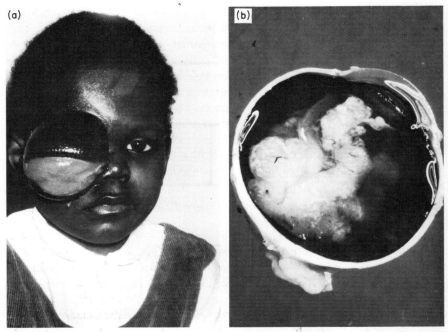

Figure 3.2:   (a) Retinoblastoma and (b) histology from the same tumor. Figure courtesy of the MRC Clinical Oncology and Therapeutic Unit, Cambridge, UK.

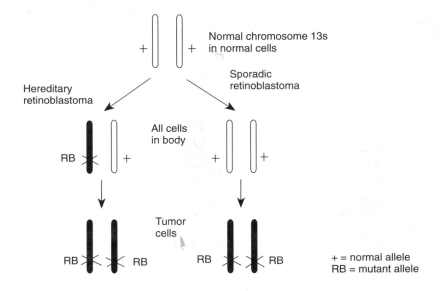

Figure 3.3:   The origin of hereditary and sporadic retinoblastomas.

region of band q14 of chromosome 13 (*Figure 3.4*) and it was inferred that the retinoblastoma gene (*RB1*) lay in this region [3]. Studies of the polymorphic enzyme, esterase D, which had previously been mapped to 13q14, supported the evidence for *RB1* in this area. In families with hereditary retinoblastoma, close linkage was demonstrated between esterase D alleles and retinoblastoma, suggesting that the two genes had to be close. In nonhereditary cases, approximately 20% of retinoblastomas but not the constitutional DNA of the patient were shown to have an abnormality; usually absence or a deletion in one copy of chromosome 13, and reduced levels of esterase D were detected in the tumor [6]. Patients with two detectable variants of esterase D in somatic tissues had only one variant present in their tumors. These studies suggested that in tumors there was loss or deletion of part of chromosome 13 and it was assumed that there was a mutation in the *RB1* gene on the remaining copy of chromosome 13.

These changes were confirmed at the molecular level by Cavanee and colleagues [7] using the loss of heterozygosity test which has become widely used for the detection of tumor suppressor genes. The loss of heterozygosity test depends on differences in the lengths of DNA fragments generated by digestion of genomic DNA with restriction enzymes. These restriction fragment-length polymorphisms (RFLPs) present within the population can be detected by DNA probes specific for the DNA fragment of interest. With the development of PCR, the technique more commonly uses microsatellite markers which are distributed throughout the genome and which are more useful than RFLP analysis because of the greater number of alleles present at any one locus. In the case of retinoblastoma, RFLP analysis was used and the probes selected for use were located on 13q. Patients suitable for study were those who had differ-

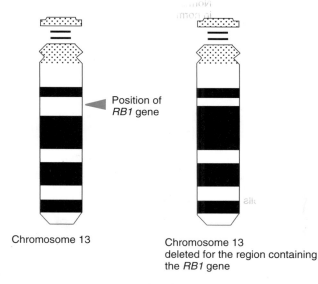

Chromosome 13

Chromosome 13
deleted for the region containing
the *RB1* gene

Position of
RB1 gene

Figure 3.4:   Deletion of 13q14 as seen in retinoblastomas.

ent sized fragments of DNA (alleles) on each of the two chromosome 13s in their somatic tissue, that is, they were heterozygous. When tumors from the same patients were analyzed, only one of the two alleles was present (*Figure 3.5*). This loss of heterozygosity (LOH) can occur by a number of possible mechanisms (*Figure 3.6*), including loss of the normal chromosome possibly followed by reduplication of the abnormal one, an interstitial deletion of the normal chromosome, or a recombination event resulting in two copies of the deficient allele. Most of the mechanisms shown in *Figure 3.6* result in loss of heterozygosity along the majority of the chromosome [7].

Narrowing the region containing the *RB1* gene made it feasible to clone the gene and study the mutations at a molecular level. Independently, three groups isolated the DNA segment which had the properties of the *RB1* gene, using basically similar approaches [8–10]. This process of identifying a gene without knowing anything about its function, product or even its location in the genome is termed 'positional cloning' and is shown schematically in *Figure 3.7* for *RB1*. A DNA sequence (H3.8) was identified which had been shown to detect deletions in a number of retinoblastomas. Based on the assumption that this

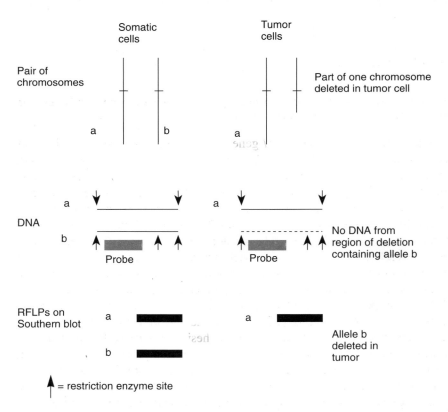

Figure 3.5:   RFLP analysis as a means of detecting loss of heterozygosity.

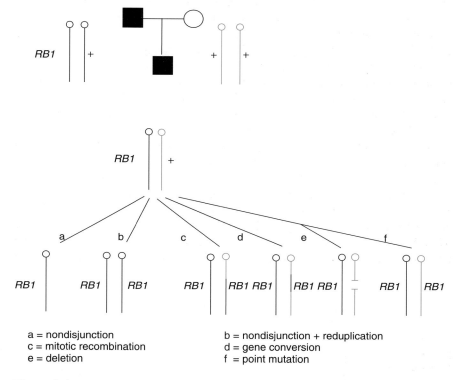

a = nondisjunction
c = mitotic recombination
e = deletion

b = nondisjunction + reduplication
d = gene conversion
f = point mutation

Figure 3.6: Mechanism of loss of heterozygosity. Reproduced with permission from *Nature* **305**, 779. © 1983. Macmillan Magazines Ltd.

sequence must lie close to the *RB1* gene, a number of chromosome walks were carried out and fragments of DNA present in single copies were identified. These were used to probe DNA from humans and other species and one sequence was found to be conserved across species suggesting that it was perhaps a coding exon of a gene. Hybridization of this sequence to RNA from retinal cell lines showed that it recognized a 4.7 kb mRNA. Finally, the conserved sequence was used to isolate a cDNA clone containing the *RB1* gene. There were several lines of evidence which verified that this was the correct sequence: (1) structural aberrations, including deletions, were seen in some retinoblastomas and osteosarcomas, (2) absence or abnormal expression of the RNA transcript was detected in tumors, and (3) a structural change in the fibroblasts of a patient with bilateral disease and a second mutation in the tumor was detected, confirming Knudson's hypothesis. The native RB protein was isolated using antisera raised against a synthetic fusion protein, which had been produced by inserting part of the *RB1* cDNA into an expression vector .

The tumor suppressor nature of RB was shown by the introduction of a single copy of *RB1* into tumor cell lines lacking the gene resulting in complete or partial suppression of the tumorigenic phenotype [11]

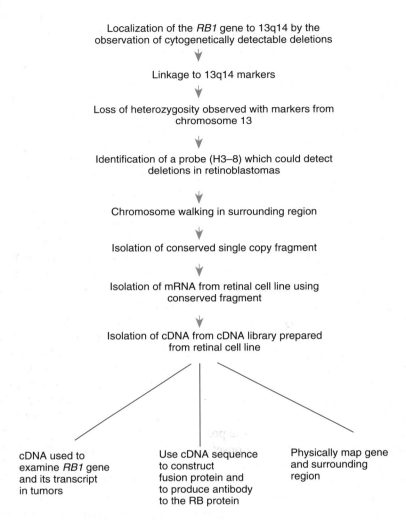

Localization of the *RB1* gene to 13q14 by the
observation of cytogenetically detectable deletions

Linkage to 13q14 markers

Loss of heterozygosity observed with markers from
chromosome 13

Identification of a probe (H3–8) which could detect
deletions in retinoblastomas

Chromosome walking in surrounding region

Isolation of conserved single copy fragment

Isolation of mRNA from retinal cell line using
conserved fragment

Isolation of cDNA from cDNA library prepared
from retinal cell line

cDNA used to
examine *RB1* gene
and its transcript
in tumors

Use cDNA sequence
to construct
fusion protein and
to produce antibody
to the RB protein

Physically map gene
and surrounding
region

Figure 3.7:   Positional cloning and identification of *RB1*.

The *RB1* gene spans approximately 180 kb of DNA and is made up of one
large exon and 26 small exons [12]. It encodes a 105 kDa nuclear phosphopro-
tein with DNA-binding activity. The protein has a regulatory role in cell proliferation, acting via transcription factors to prevent the transcriptional activation
of a variety of genes, the products of which are required for the onset of DNA
synthesis, the S phase, of the cell cycle. The first clues as to how it might work
came from studies on the interaction of viral proteins with RB. For viruses to
replicate they must force cells to undergo division by removing any barriers to
proliferation. Viral proteins have been shown to bind to RB, thereby inactivating it and allowing cell division to occur. This binding occurs in a region of the

protein called the pocket, the same region which is frequently affected in human tumors by mutations or deletions resulting in loss of RB function.

RB function is regulated by phosphorylation in a cell cycle-specific manner. It is hypophosphorylated in the G_0/G_1 phase of the cycle, hyperphosphorylated prior to G_1/S transition and throughout S, in at least a two-step process, and dephosphorylated as the cell leaves mitosis and goes back to G_0 or G_1 phase [13]. The phosphorylation state of the RB protein therefore varies throughout the cell cycle and correlates with its capacity to interact with the transcription factors described below as well as with viral oncoproteins [14]. Phosphorylation of RB occurs at multiple serine and threonine sites recognized by the cyclin-dependent kinases (CDKs), complexed with the cyclins. Phosphorylation of RB occurs through the action of cyclin D1–3/CDK4–6 in early G_1, by cyclin E/CDK2 in late G_1 and early S and by cyclin A/CDK2 in S phase. Phosphorylation of RB controls not only the ability of the protein to bind to effector proteins but also its affinity for nuclear association. In the underphosphorylated, active state, RB is tightly bound to the nuclear matrix and can bind transcription factors and viral proteins. When fully phosphorylated RB shows loose nuclear tethering characteristic of inactive RB. Mutant RB protein similarly shows loss of nuclear tethering.

RB does not act directly to control transcription but does so by interaction with transcription factors, many of which have now been identified. The best studied of these are the members of the E2F family (E2F-1–5), a family of transcription factors containing a basic helix–loop–helix (bHLH) motif. E2F-1 interacts specifically with underphosphorylated RB protein in the same region as that involved in binding viral oncoproteins. Interaction of RB and E2F-1 is stabilized by binding DP-1, a further E2F related protein. E2F binding sequences have been identified in the promoters of a variety of genes the products of which are required for the onset of DNA synthesis in S phase such as dihydrofolate reductase and thymidine kinase [15]. Interaction of RB with E2F-1 inhibits transcription of these genes and therefore suppresses cell growth. A model for the control of RB function is shown in *Figure 3.8*. Once RB is phosphorylated, E2F-1 is released allowing it to initiate transcription. The majority of RB mutations result in truncated unstable proteins which leave E2F-1 free to continually initiate transcription. Interaction of viral oncoproteins with RB also inactivate it, similarly leaving E2F-1 free and promoting cell growth. However the picture of transcription regulation is made more complex by the finding that a number of RB related proteins, (e.g. p107 and p130) also bind to members of the E2F family and are therefore also involved in regulatory processes. For example p107 has been shown to repress *MYC* activated transcription [16].

Interestingly, RB can also positively regulate transcription as has been demonstrated for the TGFβ2 gene. Here activation of transcription involves interaction of RB with the ATF2 transcription factor. In specific cell lines such as lung epithelial cell lines, the TGFβ1 gene is also activated by RB but in lines

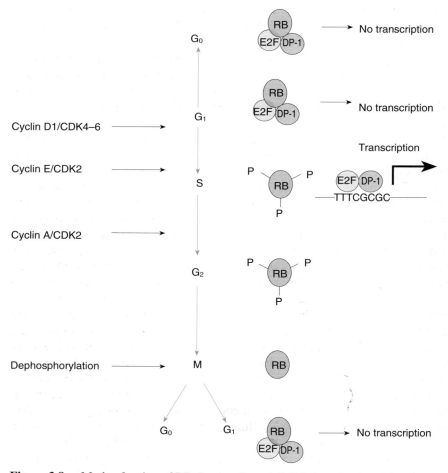

Figure 3.8:   Mode of action of RB. During G_0 and G_1 RB is underphosphorylated and is bound to the E2F-1 transcription factor complexed with DP-1. During G_1, cyclin D1/CDK4-6 and cyclin E/CDK2 phosphorylate RB and E2F-1 is released to interact with and promote transcription from genes necessary for S phase. Phosphorylation of RB is maintained by cyclin A/CDK2 until mitosis when it is dephosphorylated ready either to re-enter G_1 or to go into the stationary phase.

from different origins it is repressed, suggesting that RB regulates transcription in a cell type-dependent manner. Similarly RB has also been shown to repress transcription of the *FOS* proto-oncogene in NIH3T3 fibroblasts but promote transcription in lung epithelial lines. Recently, RB was also shown to stimulate *MYC* expression in a cell type-specific manner [17]

RB1 has been found to be mutated in a wide range of tumors and clinical aspects of mutations are discussed more fully in Chapter 11. Some of these changes, such as those seen in osteosarcomas, were expected as these tumors frequently arise in retinoblastoma patients later in life. The unexpected observation is that mutations in the *RB1* gene can be detected in unrelated tumors including breast, colon and lung. This suggests that the *RB1* gene may play an important general role in tumorigenesis.

To study the role of RB further, mice heterozygous and homozygous for mutations in *RB1* have been produced by homologous recombination. Mice carrying heterozygous mutations did not develop retinoblastomas but instead developed pituitary tumors at high frequency. These tumors are not found in humans carrying *RB1* mutations, suggesting that human and mouse tissues vary in their susceptibility to loss of RB [15]. Mice homozygous for *RB1* mutations die around 14 days of gestation as a result of failure to produce mature erythrocytes in the liver, accompanied by extensive cell death in the central nervous system [15]. In addition, the development of the lens of the eye is disrupted with a failure of differentiation. To try to get round the problems of early death in these animals, which prevented study of tissue development after 14 days, chimeric mice have been produced which have both wild-type cells and cells containing only mutant *RB1* alleles. These mice developed normally even though the homozygous mutation was present in a proportion of all tissues examined. Tumors were only found in the pituitary and aberrant cells were seen in some tissues and in particular, cells characteristic of apoptotic cells were seen in the lens and retina. The failure of erythropoiesis seen in the mice homozygous for *RB1* mutations was not seen in the chimeric mice. These results suggest that RB is essential in some tissues to regulate cell growth and its absence results in failure to produce terminally differentiated cells. However in many cell types its absence does not generally cause tumorigenesis. For tumors to develop, mutations in other genes are required. Preliminary data have shown that mice carrying mutations in both copies of *p53* and heterozygous for *RB1* mutations develop a wide range of tumors. The suggestion is that the presence of the p53 protein results in apoptosis and prevents tumorigenesis. This could be explained by a number of mechanisms: (1) either the two genes negatively regulate cell growth so that if both are lost the affect is additive; (2) loss of *p53* increases the mutation rate in *RB1*; or (3) if *p53* is absent apoptosis is prevented, allowing the *RB1* mutations to cause tumors [18].

The mode of action of RB is now being extensively studied and perhaps more questions than answers are being raised. However the 'two hit' model for *RB1* and studies on loss of heterozygosity have been central to the work being carried out in further tumor suppressor genes in other hereditary cancers.

3.3   Hereditary cancers

3.3.1   Wilms' tumor

Wilms' tumor is a rare childhood renal tumor with both a hereditary and a non-hereditary form, the former constituting only about 1% of cases. Bilateral tumors occur in approximately 10% of cases suggesting that the majority of bilateral tumors occur as *de novo* germ-line mutations rather than being inherited from an affected parent. Wilms' tumor is also associated with a complex of other conditions including aniridia, genito-urinary abnormalities and mental retardation, hence the acronym WAGR. Cytogenetic analysis of patients with WAGR syndrome showed deletions of chromosome 11 centered around 11p13, a finding which was helpful for the localization of the gene [19].

Comparison of constitutional deletions in patients with WAGR narrowed the critical region containing the gene to between the catalase gene and the gene for the follicle-stimulating hormone β subunit at 11p13. Many DNA markers were generated in this region, one of which showed homozygous deletion in a Wilms' tumor. The extent of the deletion was studied by pulsed field gel electrophoresis, then candidate genes in the area which showed evolutionary conservation and were expressed in the appropriate tissues, were studied. Finally the gene for Wilms' tumor, *WT1*, was isolated at the centromeric end of the deletion and the aniridia gene, *PAX6*, was subsequently identified at the more telomeric end. The *WT1* gene has 10 exons and encodes a transcription factor with four zinc finger DNA binding domains plus a transactivation domain in the proline/glutamine-rich amino terminal [20]. The WT1 protein shows a high degree of homology with the early growth response gene 1 (*EGR1*), a gene which is involved in the control of cell proliferation. WT1 has been shown to bind to the *EGR1* consensus binding sequences and it has been suggested that the WT1 protein may be a repressor of those genes which contain this sequence in their promoters. *WT1* undergoes alternate splicing to produce four different mRNAs which encode four different proteins which may have different functions. *WT1* is expressed in tissues of the developing kidney, the genital ridge, fetal gonad and mesothelium and is therefore involved in genito-urinary development [21]. In mouse models, homozygous inactivation of *WT1* leads to the complete absence of kidney and gonadal development but mice hemizygous for *WT1* show no such abnormality in development, nor the presence of tumors, suggesting differences in mouse and human pathways regulated by *WT1*.

Intragenic microdeletions and point mutations have been found in both sporadic and hereditary Wilms' tumors. Mutations in the zinc finger domains abolish DNA binding activity whereas mutations in the transactivation domain produce a protein which shows transcriptional activation rather than repression. There is considerable variation in the phenotype associated with *WT1* mutations. Mutation or loss of one *WT1* allele confers different degrees of genito-urinary malformations. Deletion of the entire *WT1* gene, as found in WAGR syndrome, can result in relatively mild genital anomalies whereas mutations in the third zinc finger domain can lead to the much more severe condition known as Denys–Drash syndrome (DDS). In this condition, gonadal dysgenesis, nephropathy and Wilms' tumor all occur [22]. It has been suggested that the severity of this phenotype may be because mutant WT1 forms dimers with wild-type protein and therefore acts in a dominant negative fashion, resulting in loss or reduction of DNA binding [23].

The exact role played by *WT1* in tumorigenesis is not completely clear. It has been shown that a Wilms' tumor cell line can no longer produce tumors in nude mice following the introduction of a normal chromosome 11 by microcell fusion, confirming the suppressive nature of the gene. In agreement with this, LOH has been shown to occur in both sporadic and hereditary Wilms' tumor as well as in tumors from DDS patients, suggesting that loss of the remaining allele confers additional growth advantage to the tumor. However both Wilms' tumor patients and DDS patients have been described in whom only one allele is mutated in the tumor. It therefore appears that although mutation in both alleles is one pathway to tumor formation it is not essential [23]. One possibility for this observation is that a second mutation may have occurred at a WT locus other than *WT1*.

In retinoblastoma, tumors arise as a result of mutations in only one gene, *RB1*. However Wilms' tumors can result from mutations in at least two other genes which complicate the 'two hit' model described for retinoblastoma. Beckwith–Wiedemann syndrome (BWS) is characterized by exomphalos, macroglossia and gigantism; it also carries an increased risk of developing tumors, especially Wilms' tumor [24]. Most cases are sporadic but families have been described in which the disease segregates as an autosomal dominant trait. This disorder has been mapped by linkage analysis to 11p15. A potential candidate gene in this region is insulin-like growth factor 2 (IGF2) [24]. This is an imprinted gene, being expressed only from the paternal allele. In Wilms' tumors showing loss of heterozygosity for 11p15 markers, over 90% show preferential loss of the maternal allele suggesting that imprinting may play a role in the development of Wilms' tumor. In BWS and in some Wilms' tumors, this imprinting pattern is relaxed and both alleles are expressed [25]. A third locus has been implicated in the development of Wilms' tumors as genetic linkage to both 11p13 and 11p15 has been excluded in two Wilms' tumor families [26].

3.3.2 *Familial adenomatous polyposis*

Familial adenomatous polyposis (FAP) is an autosomal dominantly inherited condition with an incidence of around 1 in 10 000. It accounts for 1–2% of all colorectal cancer. It is characterized clinically by the presence of hundreds to thousands of polyps throughout the colon and rectum (*Figure 3.9*) and at least one of these will become malignant from the second decade of life onwards. Failure to remove the tumor will result in the death of the patient, usually by 40 years of age. The manifestations of the disease are not limited to the colon and rectum but include upper gastro-intestinal polyps, osteomas, desmoid tumors, epidermoid cysts and more rarely thyroid tumors and hepatoblastoma. The observation of a cytogenetically detectable deletion in a patient with FAP and mental retardation led to the localization of the gene, termed the *APC*

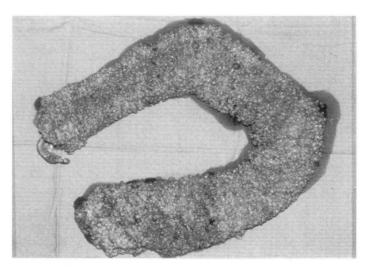

Figure 3.9:   A resected colon from a 12-year-old girl showing the presence of thousands of polyps throughout the colon.

gene (for adenomatous polyposis coli), to chromosome 5q21 by family linkage studies [27].

A positional cloning strategy was used to try to identify the causative gene for FAP. The first gene to be identified in the region was the 'mutated in colorectal cancer' gene (*MCC*). This was ruled out as the gene for FAP as no mutations were identified in it for FAP families, although it was mutated in sporadic cancers. The *APC* gene itself was identified in 1991 and was found to lie in close proximity to *MCC* [28,29]. The *APC* gene is 8.5 kb consisting of 15 exons and encoding 2843 amino acids (*Figure 3.10*). Mutations are found distributed throughout the gene and the vast majority of these introduce premature stop codons resulting in the production of a truncated protein [30]. Mutations in *APC* are discussed more fully in Chapter 6. The protein shows little homology to other known proteins and its function is just beginning to be determined by studying the various domains which can be identified along its length (*Figure 3.10*). The first third of the protein contains a series of heptad repeat sequences which are predicted to form α helical structures, capable of dimerization. Truncated proteins have been shown to form dimers with wild-type protein demonstrating the potential for dominant negative interference. The middle section of the protein contains two repeat regions through which APC has been shown to associate with the adherens junction protein, β-catenin. β-catenin, together with α- and γ-catenin, associates with E-cadherin and is essential to the integrity of the adherens junction [31]. These junctions are important for cell adhesion and intracellular communication. APC may therefore serve as a downstream target for signaling by β-catenin and may also have a role in the regulation of the β-catenin pool [31]. Finally wild-type APC has

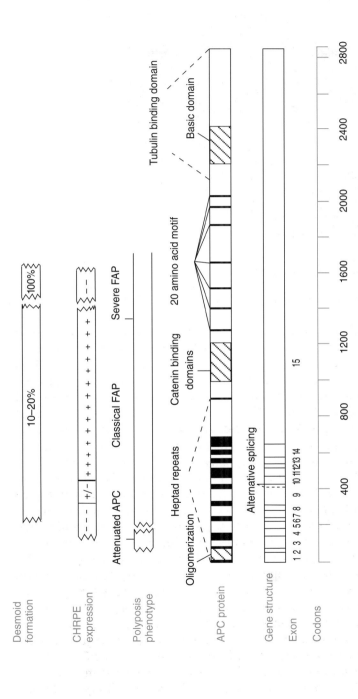

Figure 3.10: Structure of the *APC* gene showing the first 14 small exons and the large 15th exon. The amino-terminal end of the protein contains heptad repeats and it is through this region that the protein dimerizes, the first 45 amino acids being the most important. The central region of the protein contains two repeat regions through which catenin binding occurs. The carboxyl-terminal region is responsible for binding to cytoplasmic microtubules. The phenotypes associated with mutations in specific regions of the gene are also shown. Reproduced from *Hum. Molec. Genet.*, **4,** 337–340 by permission of Oxford University Press.

been shown to interact with the cytoplasmic microtubules via its carboxyl-terminal domain, a function that is lost in mutant APC proteins. The exact significance of this interaction is unknown at present but may suggest that APC is involved in the control of cell growth by serving as a link between catenins and the microtubule cytoskeleton.

APC mutations have been shown to play a role in sporadic as well as inherited colon cancer. Initially LOH was demonstrated in 50–60% of both inherited and sporadic colorectal tumors. Approximately 65% of both adenomas and carcinomas have mutations in *APC* and these have been found in adenomas as small as 3 mm, indicating that *APC* mutation is an early event in the genesis of colorectal cancer. Inactivation of both copies of *APC*, either by mutation of each allele or mutation in one copy and deletion of the second is found in both adenomas and carcinomas, indicating that *APC* is a true tumor suppressor gene [32].

3.3.3 *Neurofibromatosis 1*

Von Recklinghausen neurofibromatosis (NF1) is an autosomal dominant disorder with an incidence of 1 in 1000, characterized by *café au lait* spots, multiple neurofibromas and an increased risk of cancers such as neurofibrosarcomas and phaeochromocytomas. The expression of the disease is quite variable from one individual to the next, so much so that mild cases can even go undiagnosed. The disease was localized to chromosome 17 by linkage analysis and also by the cytogenetic observation of translocations involving 17q11. These translocations were helpful in bringing about the rapid isolation of the *NF1* gene [33,34]. The large gene spans around 350 kb of DNA, produces a transcript of 11–13 kb and has 59 exons. Interestingly, three small genes transcribed in the opposite orientation have been found in intron 27 of the *NF1* gene. It is not yet clear whether these play any part in the disease. The protein product of *NF1* is a 2818 amino acid protein called neurofibromin. As mentioned in Chapter 2, neurofibromin shares sequence homology in the central region, encoded by exons 21–27A , with the mammalian GAP proteins which play a role in regulating the *RAS* proto-oncogene. However neurofibromin is likely to have other functions, a feature indicated by the strong sequence conservation in regions outside of the area homologous to GAP.

Most mutations so far identified are predicted to disrupt or inactivate neurofibromin (see Chapter 11). Reduced GAP activity is likely to lead to increased levels of active RAS p21 resulting in abnormal signaling through pathways described in Section 2.2.1 [see ref. 8 in Chapter 2]. However there are complications with this simple model, for example loss of neurofibromin occurs in neuroblastomas without accumulation of active RAS, which means that further research into the mode of action of neurofibromin is necessary.

NF1 is generally considered to be a tumor suppressor gene since its inactivation is associated with cell proliferation, in part involving the *RAS* gene. LOH on chromosome 17 has been found in neurofibrosarcomas with mutations in the retained allele [35]. However by no means all tumors show LOH and in some cases, those that did show loss showed a deletion on 17p rather than 17q. LOH affecting the *p53* gene has been shown to occur in some neurofibrosarcomas. Some recent studies have been carried out on LOH in neurofibromas and have indicated that somatic deletions of *NF1* are present in some benign lesions [36].

3.3.4   *Neurofibromatosis 2*

Neurofibromatosis type 2 (NF2) is much less common than NF1 with an incidence of 1:35 000–1:40 000. It is characterized by bilateral vestibular schwannomas and there is a predisposition to other tumors such as meningiomas, astrocytomas and spinal schwannomas [37]. The gene for the disorder was mapped to chromosome 22q prompted by LOH at this position in vestibular schwannomas. Sporadic forms of the disease often show monosomy for chromosome 22.

The gene for NF2 has now been cloned and has been called both schwannomin or merlin. It is a much smaller gene than that for NF1 with 17 exons one of which is alternately spliced and encodes a protein of 595 amino acids. Schwannomin is homologous to proteins which link the cell membrane with the cytoskeleton but its exact function remains unknown. Many mutations have been found in *NF2*, the majority of which are chain terminating (see Chapter 11). The tumor suppressor nature of the gene is confirmed by the many studies which have shown LOH on 22q associated with an inactivating mutation in the other copy of *NF2* both in sporadic schwannomas and meningiomas as well as in familial cases. Recent studies have shown a high rate of mutations in tumors other than those associated with NF2 such as malignant mesothelioma.

3.3.5   **BRCA1** *and* **BRCA 2**

Breast cancer is the most common malignancy in women with around 1 in 12 women affected by age 85. Hereditary breast cancer, inherited in an autosomal dominant manner, accounts for around 5% of all cases. In site-specific breast cancer families, breast cancer alone is seen whereas in breast–ovarian cancer families women are at increased risk of both cancers. Male breast cancer is also seen in some families with these conditions.

Familial breast cancer was linked to a gene on chromosome 17q, called *BRCA1*, in 1990 in families primarily with early onset breast cancer [38]. However, it was clear from the outset that more than one gene was involved in the two forms of hereditary breast cancer; breast–ovarian cancer families particularly showed linkage to the gene on chromosome 17 while those with breast cancer only were more likely to be linked to a second gene, *BRCA2*.

Positional cloning eventually led to the isolation *BRCA1* in 1994 [39].

BRCA1 spans more than 80 kb of DNA and encodes a 7.8 kb transcript composed of 24 exons. The protein is an 1863 amino acid RING finger protein. Mutations have been identified throughout the gene and are primarily chain terminating [40]. The exact function of BRCA1 is still to be determined but recent evidence shows that it has some sequence homology to the granins. These proteins are secreted and in some cases their expression is regulated by estrogen. Their exact function is not clear but they may assist in the packaging of peptide hormones and/or modulate their processing [41]. Recent studies have suggested that they may play a role in mammary epithelial cell growth. Although widely expressed in many tissues, it does appear that BRCA1 is not essential for normal life. A woman homozygous for a *BRCA1* mutation in her germ-line has been described who has breast cancer but is otherwise well [40].

Like the genes described above for other hereditary cancers, *BRCA1* also appears to be a tumor suppressor gene with LOH of the wild-type allele being a common feature of both breast and ovarian tumors in people carrying germline mutations. Where *BRCA1* differs from *RB1*, *WT1* and *APC* is that mutations have not been identified so far in the corresponding sporadic form of the disease although a small number of ovarian cancers have been identified as having *BRCA1* mutations [40].

BRCA2 was mapped by linkage analysis to 13q12–13. This gene is thought to confer a lower risk of ovarian cancer than *BRCA1* but may account for as much as 15% of all male breast cancer. *BRCA2* has recently been cloned but the function of this gene remains to be determined. Preliminary evidence shows no similarity to other genes so far identified except a very slight resemblance to *BRCA1* [42].

3.3.6   *Other hereditary cancers associated with tumor suppressor genes*

Other diseases associated with cancers and which show a hereditary form are listed in *Table 3.1*.

Table 3.1:   Loss of heterozygosity in inherited cancers

Disease	Tumor	Site of gene	LOH[a]
MEN1	Anterior pituitary insulinomas	11q12–13	11
Von Hippel–Lindau	Renal cell cancer	3p25–26	3p
Gorlins syndrome	Basal cell carcinomas Medulloblastoma	9q	
Tuberous sclerosis (TSC1)		9q34	9q
Tuberous sclerosis (TSC2)		16p13.3	16p

[a]LOH, loss of heterozygosity.

3.4 p53

The *p53* gene is located on the short arm of chromosome 17. Much is already known about the gene since its identification over 10 years ago primarily because of its involvement in a wide range and number of tumors and its important role in tumorigenesis (*Table 3.2*) [43].

p53 was originally thought to be a tumor antigen as it was found complexed to the SV40 large T antigen. Subsequently mutant *p53* was shown to cooperate with *HRAS* to transform cells and was therefore believed to be an oncogene. However its ability to abolish the tumorigenic phenotype if transfected into tumor cell lines, the frequent finding that both alleles were inactivated in tumor cells and its association with hereditary cancers led to its reclassification as a tumor suppressor gene.

p53 functions as a cell cycle checkpoint protein by transactivation of genes which encode proteins with growth suppressing activities and it exerts its function during the G_1 phase of the cycle. However *p53*, like *RB1*, is not necessary for normal cell growth and function as it has been shown that transgenic mice with deletions of both copies of *p53* develop normally. *p53* is necessary nevertheless to suppress tumor development as these mice develop tumors after a few months.

A major role for *p53* has been termed 'the guardian of the genome'. There is a rapid increase in the level of p53 in response to DNA damage which causes arrest of the cell cycle during G_1. This gives the cell time to repair its DNA. If repair is not possible, *p53* induces programed cell death, a process termed apoptosis [44]. It has recently been shown that in the center of large tumors, there is a lack of oxygen, hypoxia, which is also capable of induction of *p53* without the necessity for DNA damage. This process results in growth arrest or apoptosis [45]. Cells carrying mutations in *p53* are able to resist this process as described below and can therefore become the predominant cells in the tumor. A related role for *p53* is its ability to prevent or repair gene amplification which can arise due to genetic instability. Now that the structure of the *p53* gene and its protein product are well characterized, and the proteins with which it interacts identified, the pathways by which *p53* causes growth suppression and apoptosis have been defined.

Table 3.2: Tumors with deletions of 17p

Adrenal cortical tumors	Lung tumors
Bladder cancer	Neurofibrosarcomas
Brain tumors	Osteosarcomas
Breast cancer	Ovarian cancer
Cervical cancer	Renal cell cancer
Colorectal tumors	Testicular tumors
Hepatocellular cancer	

3.4.1   The p53 *gene and its protein product*

p53 is made up of 11 exons and encodes a 2.8 kb mRNA. The protein product of *p53* is virtually undetectable in normal cells and has a half-life of around 20 min. It is primarily located in the nucleus but can be detected in the cytoplasm in G_1 and also following DNA synthesis. The protein product of *p53* contains three distinct regions with differing functions (*Figure 3.11*). The amino terminus contains many acidic amino acids and a large number of prolines. This domain is responsible for the transactivational properties of p53. The carboxyl terminus is hydrophilic and has many highly charged residues. This region contains three nuclear localization signals and mutations in this region can lead to a protein which is located primarily in the cytoplasm. This domain is also responsible for the ability of p53 to form oligomers. Recent NMR studies have described the three-dimensional structure of this region and have shown that two carboxyl-terminal peptides form a tight dimer and in turn these dimer pairs assemble into tetramers [44]. The central region of p53 is highly hydrophobic. It is highly conserved between species and the majority of mutations in human tumors are found here (*Figure 3.11*). It is this region which shows DNA binding activity and which interacts with the target sequences in the genes which undergo transcriptional transactivation. X-ray crystallography has shown that this region is folded into β sheets to form a sandwich with protruding residues which make contact with the major and minor grooves in the DNA to which it

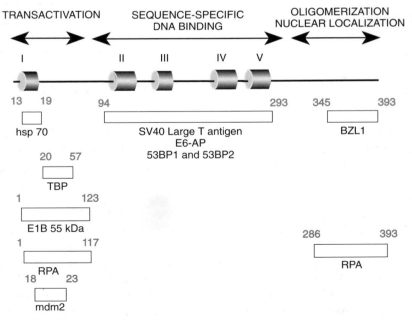

Figure 3.11:   Structure of the *p53* gene showing the five highly conserved regions. The regions at which protein partners bind to p53 are also indicated.

binds [44]. Mutations in this region either interfere with the protein–DNA interaction or affect the stability of the 3D structure.

3.4.2   p53 as a transcriptional transactivator

p53 inhibits cell growth by acting as a sequence-specific transcription factor, promoting the transcription of genes which encode proteins with growth suppressive activity. A consensus binding sequence has now been identified and shown to be a 10 bp motif which is repeated. In 1993, a gene was identified which is specifically induced by an excess of wild-type p53 [44]. This gene, which contains the consensus binding motif, was called *WAF1* and encodes the p21 protein. Overexpression of p21 can lead directly to growth suppression. *WAF1* was subsequently shown to be identical to a gene, termed *CIP1*, isolated because of its ability to bind to the cyclin-dependent kinase, CDK2. Activation of *p53* results in the transactivation of *WAF1* and in turn leads to inactivation of CDK–cyclin complexes which block cell cycle progression and result in growth arrest. As described in the section on *RB1*, the CDK–cyclin complexes are responsible for the phosphorylation of RB at different stages of progression through the cell cycle. The next step of the pathway therefore suggests that inhibition of CDK–cyclin complexes by p21 prevents the phosphorylation of RB, a step which is required for progression of the S phase of the cell cycle (*Figure 3.12*). Degradation of p53 would lead to a reduction in WAF1 levels, phosphorylation of RB and continuation through the S phase of the cell cycle. In addition, wild-type p53 can enhance transcription from the *RB1* promotor and so may also aid growth arrest directly by increasing the levels of RB protein. Inactivation of *p53*, leading to cell proliferation can occur via mutation or interaction with a number of proteins described in the next section.

3.4.3   Inactivation of p53 and its interactions with protein partners

At least 17 proteins have been shown to interact with p53 in different regions of the protein (*Figure 3.11*). Products of transforming viruses such as SV40 and human papilloma virus (HPV) interact with the central domain leading to inactivation of p53 resulting in cell proliferation and allowing viral multiplication to occur.

An interesting protein found to interact with the amino terminal end of p53 is the p90 protein which is now known to be the product of the cellular oncogene *MDM2* (mouse double minute 2). Interaction of this protein with p53 decreases transcriptional transactivation. Amplified *MDM2* was originally described in human sarcomas with wild-type *p53* and its overexpression was believed to be the mechanism by which it caused inactivation of *p53* [46].

The carboxyl-terminus of p53 associates with the DNA replication factor RPA. This association inhibits the single-stranded DNA binding activity

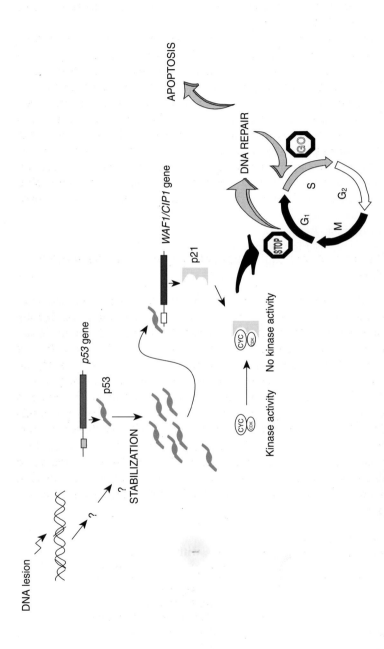

Figure 3.12: Mode of action of p53. DNA damage results in stabilization of p53. This results in transactivation of *WAF1*. The p21 protein product then complexes with cyclin/CDKs to inhibit their kinase activity. In turn this prevents phosphorylation of RB which will stop the release of E2F-1 transcription factor and maintains the cell in G$_1$ until DNA is repaired and the cell cycle can proceed. In some cases the cycle does not resume and the cell dies by apoptosis.

of RPA and may therefore be the mechanism by which *p53* can negatively regulate DNA replication. This mechanism of growth control is however likely to be more complex as mutant *p53*, which has no growth suppressive activity, also binds to RPA.

The heat shock protein hsp70 interacts with the amino terminus of mutant, though not wild-type, p53 and may be responsible for its stabilization.

Records of over 3400 mutations in *p53* are now stored in the EMBL database. Around 95% of mutations in *p53* are found in the central domain of the protein. In the majority of cases these mutations result in a protein which has lost its transactivational activity. Many of the *p53* mutations affect a CpG dinucleotide, suggesting that the most common cause of mutations is the spontaneous deamination of 5-methylcytosine resulting in C to T transitions. Mutations in *p53* are found in the majority of human cancers and are frequently found in conjunction with loss of the wild-type allele. The relationship between these mutations and clinical features is discussed in subsequent chapters.

3.5   Evidence for tumor suppressor genes in sporadic cancers

Comparison of allele loss in hereditary cancers with that seen in sporadic cancers, for example, FAP and sporadic colorectal cancer, has suggested that the fundamental mechanism for carcinogenesis may be the same for both. Many sporadic cancers have been examined for loss of heterozygosity and *Table 3.3* gives an overview of those tumors in which allele loss has been demonstrated. It is still not clear whether all these regions contain tumor suppressor genes.

It is apparent that in many tumors more than one chromosomal region is involved. This is only to be expected if we imagine tumors arise via a multistage process involving several genes. *Table 3.3* also shows that there is no particular tissue specificity for the tumor suppressor genes. Loss of 17p, presumably involving *p53*, has, for example, been implicated in a wide range of tumors and the prototype tumor suppressor *RB1* has been mutated in such diverse tumors as small cell lung cancer (SCLC), breast cancer and hematological malignancies.

Table 3.3:   Allele loss in human tumors

Tumor	Chromosomal region lost
Bladder	2p, 3p, 11p
Breast	1q, 3p, 11p, 13q, 17p, 17q
Colon	On all except chromosome 2
Lung	3p, 11p, 13q, 17p
Ovary	11p
Stomach	13, 18q
Testicular	11p, 17p

3.6   Interaction and differences between oncogenes and tumor suppressor genes

As mentioned in Chapter 2 and alluded to at various points in this chapter, no single genetic event causes tumors. Even retinoblastoma, which has been described as being caused by mutations in a single gene, does not contradict this because at least two hits are required in the gene for the tumor to develop. Other oncogenes, for example, *NMYC*, have also been shown to contribute to the development of retinoblastomas. Colorectal cancer has been studied perhaps more than any other tumor when investigating the ways in which tumor suppressor genes and oncogenes might interact, and as such serves as a good model for other cancers.

3.6.1   *Interaction between oncogenes and suppressor genes – the colorectal cancer model*

Colorectal cancer has several hereditary forms which have given some clues as to the genes involved in the development of sporadic forms of the cancer (see Chapter 6). It also has a well-defined pattern of progression from adenoma to carcinoma. For these reasons, and also because it is one of the commonest forms of cancer, more is known about colorectal cancer than most other sporadic cancers. Table 3.4 shows the major changes in oncogenes and tumor suppressor genes seen in colorectal cancer [47]. RAS has been discussed in Chapter 2 and the APC gene and p53 in this chapter. The DCC gene (deleted in colorectal cancer) has recently been identified following the observation of a high rate of allele loss on chromosome 18q which suggested the existence of a tumor suppressor. The DCC gene encodes a protein which is similar in sequence to neural cell adhesion molecules and other related cell surface gly-coproteins and is believed to participate in signaling pathways which regulate cell proliferation. It is present in most normal tissues but its expression is reduced or absent in the majority of colorectal cancers. Its role in the development of colorectal cancers is probably via alterations in the regulation of cell-to-cell contact [48] (Table 3.4).

Table 3.4:   Involvement of oncogenes and tumor suppressor genes in sporadic colorectal cancer

Gene	Location	Percentage of adenomas			Percentage of cancers
		Early	Intermediate	Late	
APC	5q21	60	60	60	60
DCC	18q	13	11	47	70
KRAS	12q	10	50	50	50
p53	17p	6	6	24	75

Data from ref. 46 and personal communications.

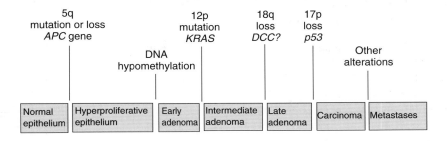

Figure 3.13: Model for the interaction of oncogenes and tumor suppressor genes in colorectal tumorigenesis. Reproduced from ref. 47 with permission from Cell Press.

These alterations have been put together to suggest a model for tumorigenesis as shown in *Figure 3.13* [47]. Other oncogenes, (e.g. *MYC*), tumor suppressor genes (e.g. *RB1*), and loss of other chromosome regions such as 1q, 4p, 6p, 6q, 8p, 9q and 22q have also been seen in colorectal cancers and may be involved at various stages. In addition, other epigenetic phenomena such as loss of methyl groups have been shown to occur early in tumorigenesis and may contribute to the instability of the genome. This colorectal model is similar to that described for other tumors such as SCLC, breast cancer and melanoma where, again, preliminary evidence for interactions between oncogenes and tumor suppressor genes has been observed. Such a model provides us with a framework on which to expand.

References

1.   Harris, H. (1988) *Cancer Res.*, **48**, 3302.
2.   Stanbridge, E.J. (1986) *BioEssays*, **3**, 252.
3.   Horsthemke, B. (1992) *Cancer Genet. Cytogenet.*, **63**, 1.
4.   Knudson, A.G. Jr. (1971) *Proc. Natl Acad. Sci. USA*, **68**, 820.
5.   Knudson, A.G.(1993) *Proc. Natl Acad. Sci. USA*, **90**, 10914.
6.   Benedict, W.F., Banerjee, A., Mark, C. and Murphree, A.L. (1982) *Cancer Genet. Cytogenet.*, **6**, 213.
7.   Cavenee, W.K., Dryja, T.P., Phillips, R.A., Benedict, W.F., Godbout, R., Gallie, B.L., Murphree, A.L., Strong, L.C. and White, R.L. (1983) *Nature*, **305**, 779.
8.   Friend, S.H., Bernards, R., Rogelj, S., Weinberg, R.A., Rapaport, J.M., Albert, D.M. and Dryja, T.P. (1986) *Nature*, **323**, 643.
9.   Lee, W.-H., Bookstein, R., Hong, F., Yoing, L.J., Shew, J.H. and Lee, E.Y.-H.P. (1987) *Science*, **235**, 1394.
10.   Fung, Y.K.T., Murphree, A.L., T'ang, A., Qian, J., Hinrichs, H.S. and Benedict, W.F. (1987) *Science*, **236**, 1657.
11.   Huang, H.-J.S., Yee, J.-K., Shew, J.-L., Chen, P.-L., Brookstein, R., Friedmann, T., Lee, E.Y.-H.P., Lee, W.-H. (1988) *Science*, **252**, 1563.
12.   Toguchida, J., McGee, T.L., Paterson, J.C., Eagle, J.R., Tucker, S., Yandell, D.W. and Dryja, T.P. (1993) *Genomics*, **16**, 535.

13. Buchkovich, K., Duffy, L.A. and Harlow, E. (1989) *Cell*, **58**, 1097.
14. Mittnacht, S., Lees, J.A., Desai, D., Harlow, E., Morgan, D.O, and Weinberg, R.A. (1994) *EMBO J.* **13**, 118.
15. Hinds, P.W. (1995) *Curr. Opin. Genet. Dev.*, **5**, 79.
16. Gu, W., Bhatia, K., Magrath, I.T., Dang, C.V. and Dalla-Favera, R. (1994) *Science*, **264**, 251.
17. Adnane, J. and Roberts, P.D. (1995) *Oncogene*, **10**, 381.
18. Williams, B.O., Remington, L., Albert, D.M., Mukai, S., Bronson, R.T. and Jacks, T. (1994) *Nature Genetics*, **7**, 480.
19. Riccardi, V.M., Sujansky, E., Smith, A.C. and Franke, U. (1978) *Paediatrics*, **61**, 604.
20. Gessler, M., Poustka, A., Cavenee, W., Neve, R.L., Orkin, S.H. and Bruns, G.A.P. (1990) *Nature*, **343**, 774.
21. Van Heyningen, V. and Hastie, N.D. (1992) *Trends Genet.* **8**,16.
22. Mueller, R. (1994) *J. Med. Genet.*, **31**, 471.
23. Little, M.H., Williamson, K.A., Mannens, M., Kelsey, A., Gosden, C., Hastie, N.D. and van Heyningen, V. (1993) *Hum. Mol. Genet.*, **2**, 259.
24. Elliot, M. and Maher, E.R. (1994) *J. Med. Genet.*, **31**, 360.
25. Weksberg, R.Shen, D., Fei, Y.L., Song, Q. and Squire, J. (1993) *Nature Genetics*, **5**, 143.
26. Grundy, P., Koufos, A., Morgan, K., Li, F.P., Meadows, A.T. and Cavenee, W.K. (1988) *Nature*, **336**, 374.
27. Bodmer, W.F., Bailey, C.J., Bodmer, J. *et al.* (1987) *Nature*, **328**, 614.
28. Groden, J., Thliveris, A., Samowitz, W. *et al.* (1991) *Cell*, **66**, 589.
29. Kinzler, K.W., Nilbert, M.C., Su, L.-K. *et al.* (1991) *Science*, **253**, 661.
30. Wallis, Y. and Macdonald, F. (1996) *J. Clin. Mol. Pathol.* **49**, M65–M73.
31. Polakis, P. (1995) *Curr. Opin. Genet. Dev.*, **5**, 66.
32. Nagase, H. and Nakamura, Y. (1993) *Hum. Mutat.*, **2**, 425.
33. Visklochil, D., Buchberg, A.M., Xu, G. *et al.* (1990) *Cell*, **62**, 187.
34. Wallace, M.R., Marchuk, D.A., Andersen, L.B. *et al.* (1990) *Science*, **249**, 181.
35. Xu, W., Mulligan, L.M., Ponder, M.A., Liu, L., Smith, B.A., Mayhew, C.G.P. and Ponder, B. (1992) *Genes Chrom. Cancer*, **4**, 337.
36. Colemand, S.D., Williams, C.A. and Wallace, M.R. (1995) *Nature Genetics*, **11**, 90.
37. Thomas, G. (1994) *Eur. J. Can.*, **30A**, 1981.
38. Hall, J.M., Lee, M.K.,Morrow, J., Newman, B., Anderson, L., Huey, B. and King, M.-C. (1990) *Science*, **250**, 1684.
39. Miki, Y., Swensen, J., Shattuck-Eidens, D. *et al.* (1994) *Science*, **266**, 66.
40. Szabo, C.I. and King, M.-C. (1995) *Hum. Mol. Genet.*, **4**, 1811.
41. Jensen, R.A., Thompson, M.E., Jetton, T.L. *et al.* (1996) *Nature Genetics*, **12**, 303.
42. Wooster, R., Neuhausen, S.L., Mangion, J. *et al.* (1994) *Science*, **265**, 2088.
43. Donehower, L.A. and Bradley, A. (1993) *Biochim. Biophys. Acta*, **1155**, 181–205.
44. Haffner, R. and Oren, M. (1995) *Curr. Opin. Genet. Dev.*, **5**, 84.
45. Kinzler, K. and.Vogelstein, B. (1996) *Nature*, **379**, 19.
46. Oliner, J.D., Kinzler, K., Melitzer, P.S., George, D.L. and Vogelstein, B.(1992) *Nature*, **358**, 80.
47. Fearon, E.R. and Vogelstein, B. (1990) *Cell*, **61**, 759.
48. Cho, K.R. and Fearon, E.R. (1995) *Curr. Opin. Genet. Dev.*, **5**, 72.

Chapter 4

Cell cycle control genes and mismatch repair genes

4.1 Introduction

In the last few years two further groups of genes have been shown to play a major role in the development of cancer and need consideration in their own right. The first group of genes are those intimately involved in the positive and negative control of the cell cycle. These interact closely with both the oncogenes and the tumor suppressor genes and can, in some cases, be classed as such themselves. The second group of genes are those initially identified in colorectal cancer which have been shown to be involved in the repair of DNA mismatches.

4.2 Cyclins and cyclin-dependent kinases

The stages of the cell cycle and the checkpoints which exist to prevent entry into subsequent stages with damaged DNA are described in Chapter 1. This progression from one stage to the next is carefully controlled by the sequential activation and degradation of the cyclins and activation of their partners, the cyclin-dependent kinases (CDKs).

Thirteen mammalian cyclins have now been identified (*Table 4.1*), each one of which is required at a different stage of the cell cycle [1]. Each cyclin undergoes a characteristic pattern of synthesis and degradation dependent on the stage of the cycle at which it acts (*Figure 4.1*). The cyclins are broadly classified into G_1 and mitotic cyclins, according to the stage of the cycle during which they are produced. They all contain a homologous region of about 100 amino acids called the cyclin box and it is this region of the protein which binds to the cyclin's partner, the CDK. The G_1 cyclins are relatively short lived proteins and are degraded, when no longer required, by a process conferred by the so-called PEST sequences, characteristic of short lived proteins, which lie C terminal to the cyclin box [2]. The mitotic cyclins are longer lived and are degraded, prior to entry into mitosis, by proteinases via a ubiquitin-dependent pathway. The instability in this case is conferred by the 'destruction' box located N terminal to the cyclin box [1]. This pattern of degradation of the cyclins is important as it necessarily inactivates the CDKs.

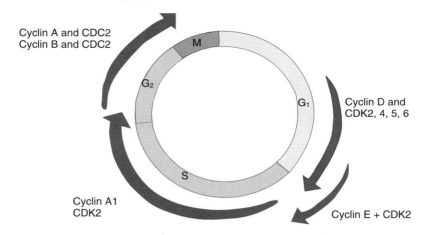

Figure 4.1:   The stages of the cell cycle and expression of the cyclins and CDKs.

Table 4.1:   Mammalian cyclins and cyclin-dependent kinases

Cyclin	Family	CDK	Stage
A	Mitotic	CDC2, CDK2	S, G_1, M
B 1–3	Mitotic	CDC2	M
C	G_1	?	?
D1–4	G_1	CDK2,4,5,6	G_1
E	G_1	CDK2,4,5,6	G_1/S
F	?	?	?
G	?	?	?
H	?	$p40^{mo15}$	?

Six mammalian CDKs have so far been identified (*Table 4.1*). The CDKs are activated by complexing with the cyclins and by a pattern of phosphorylation and dephosphorylation at specific residues on the kinases. Activation of the CDKs occurs by phosphorylation of a conserved threonine residue (at position 160) and by binding of the cyclin (*Figure 4.2*). The protein responsible for the phosphorylation of CDC2, CDK2 and CDK4 has now been identified as $p40^{mo15}$ and this in turn is activated by another cyclin, cyclin H. In addition, the action of cyclin/CDKs can be inhibited by phosphorylation, a process which has now been defined clearly for two CDKs, CDC2 and CDK2. This phosphorylation occurs on threonine 14 or tyrosine 15 and is controlled and maintained throughout interphase by the protein kinase, wee1/mik1, thereby keeping the CDK in its inactive state [1]. At the end of G_2, another phosphatase, the product of the CDC25 gene, induces dephosphorylation of these residues to activate the kinase (*Figure 4.2*).

4.3   Cyclin-dependent kinase inhibitors

Control of cyclins and CDKs is now also known to occur via a group of

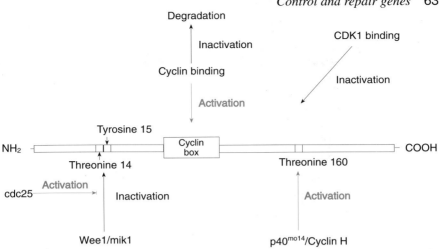

Figure 4.2: Activation and inhibition of the cyclin-dependent kinases showing the amino acids which are phosphorylated and dephosphorylated. Cyclin binds to the cyclin box, the exact position of binding of the CDKIs remains unknown at present. Steps involving activation are shown in orange and those which are inhibitory are shown in black. Adapted from ref. 1 and reproduced with permission from Academic Press Inc.

inhibitor proteins known as cyclin-dependent kinase inhibitors (CDKIs). There are seven different CDKIs in mammalian cells which belong to two different classes [3] (*Table 4.2*). The first class comprises p21, p27 and p57. These inhibitors preferentially bind to the G_1/S class of CDKs. They are classified as dual specificity inhibitors as they can bind not only to the kinases but also to other proteins. For example, p21 can interact with the DNA replication factor, proliferating cell nuclear antigen (PCNA) thereby inhibiting polδ-catalyzed DNA replication. The second class of inhibitor is the CDK4I (inhibitor of CDK4) family comprised of the ankyrin repeat proteins. These include p15, p16, p18 and p19. These inhibitors act on cyclin D complexed either to CDK4 or CDK6.

Table 4.2: Cyclin-dependent kinase inhibitors

Inhibitor	Target	Chromosomal location	Regulator
p15	CDK4,6	9p21	TGFβ
p16	CDK4,6	9p21	?
p18	CDK4,6	1p32	?
p19	CDK4,6	19p13	?
p21	CDK2,3,4,6	6p21	p53, TGFβ
p27	CDK2,4,6	12p12–13	Rapamycin cAMP
p57	CDK2,3,4	11p15	?

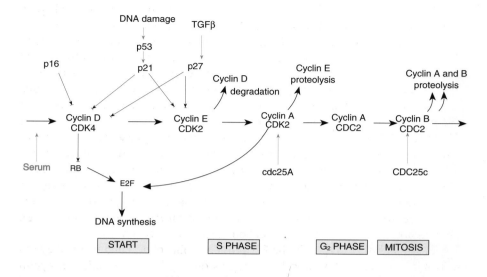

Figure 4.3:   Interaction of the cyclins/ CDKs and three of the CDKIs. Steps involving activation are shown in orange and those which are inhibitory are shown in black. Adapted from ref. 1 and reproduced with permission from Academic Press Inc.

4.4   Control of the cell cycle

Each of the cyclin–CDK complexes, together with the CDKIs, are responsible for controlling different stages of the cell cycle by preventing progression through checkpoints in the presence of DNA damage (*Figure 4.3*). Deregulation of many of these processes has now been implicated in tumorigenesis.

4.4.1   START

The cyclin–CDK complexes linked to the regulation of START are the D type cyclins [4]. There are four cyclin Ds recognized so far, termed D1, D2, D3 and D4, which are expressed in a cell lineage-specific manner. They are synthesized in response to growth factors and are very short lived. They are rapidly degraded when growth stimuli are withdrawn regardless of the position of the cell cycle and if removed during G_1, cells will not enter S phase. However degradation of D type cyclins later in the cell cycle has no effect. The D type cyclins are found in partnership with four kinases, CDK2, -4 -5 and -6 although CDK4 appears to be the main partner in most cell types (*Table 4.1*). Activation of CDK4 by complexing with cyclin D and by phosphorylation on threonine 160 drives cells through START. Deregulation of cyclin D synthesis will make cells less dependent on growth stimuli and is likely therefore to contribute to tumorigenesis.

Cyclin D1 is encoded by the *CCND1* gene on chromosome 11q13. This gene has been shown to be identical to the *BCL1* and *PRAD1* proto-oncogenes [5]. Its overexpression is associated with a number of tumors such as esophageal, breast and gastric cancers as discussed in subsequent chapters. Chromosome rearrangements involving 11q13 are found in several tumors such as parathyroid adenomas and B-cell lymphomas. In these cases, the rearrangement results in overexpression of the gene by bringing it respectively under the influence of the parathyroid promoter or the immunoglobulin heavy chain enhancer. Cyclin D2 is encoded by the *CCND2* gene on chromosome 12p13 and has been identified as the *VIN1* proto-oncogene. This gene is involved in mouse T-cell leukemia in which cyclin D2 overexpression is found and has recently been shown to be amplified in human colorectal cancer [6]. Cyclin D3 is located at 6p21 and is encoded by the *CCND3* gene but a specific role in human cancers has not yet been found although abnormalities of chromosome 6 *per se* are seen in lymphoma, acute lymphocytic leukemia and breast cancer [4].

As discussed in Chapter 2, the RB protein is phosphorylated in a cell cycle-dependent manner. Phosphorylation begins during G_1 by the action of cyclin D1/CDK4. This inactivates RB, releasing the E2F transcription factor and allows it to activate genes necessary for DNA synthesis.

CDK4 has also been shown to have a potential role in tumorigenesis and may be a target for TGFβ in some cells. The CDK4 gene has been shown to be amplified in some tumors and this may indicate a method by which cells can become insensitive to arrest by TGFβ [7].

The CDKIs are also involved in tumor development at this stage of the cell cycle. p16, encoded by the *CDKN2* or *MTS1* gene, is an inhibitor of CDK4 which acts by binding to it in competition with cyclin D [8] and has also been shown to inhibit CDK6. It can therefore prevent these kinases from phosphorylating RB (*Figure 4.3*). p16 was originally described as a CDK4 binding protein in SV40 transformed cells and has subsequently been shown to be deleted in many tumors. It is the likely candidate for the familial melanoma gene (see Chapter 11). The gene for p16 lies in tandem with the gene for p15, *MTS2*. p15 shows a high degree of homology to p16 and also functions to inhibit CDK4 and CDK6. p15 has been shown to be the protein, which together with p27, causes cell cycle arrest by TGFβ [9]. It has been suggested that p27 is found in complexes with cyclin D–CDK4. Increased expression of p15 in response to TGFβ leads to displacement of p27 from this complex [3].

4.4.2   G1 to S phase

The E type cyclins are believed to act after the D type and to be important for the initiation of DNA replication. Maximum levels of activity occur at the G_1–S transition. Cyclin E is expressed towards the end of G_1 and complexes with CDK2 to activate it. As with cyclin D–CDK4, phosphorylation of the threonine residue (160) is necessary for activation. After cells have entered S phase, cyclin E is rapidly degraded and CDK2 is released to be complexed by cyclin A at the next stage [10].

The substrates on which cyclin E–CDK2 act are not yet completely clear. It has been suggested that they may be involved in the transcription of genes necessary for S phase having been identified in complexes with the E2F transcription factor and the RB-like proteins p107 and p130 [1]. TGFβ causes a decrease in the levels of cyclin E in some cells so overexpression of this cyclin, as found in some breast cancers, may override growth inhibitory signals. There is currently no direct evidence that cyclin E can function as an oncogene [4].

Cells which have suffered DNA damage are prevented from entering S phase and are blocked at G_1. This process is dependent on the tumor suppressor gene *p53* (see Section 3.4) and the cyclin-dependent kinase inhibitor, p21. p21 is transcriptionally regulated by *p53* and provides a link between the tumor suppressors, cell cycle proteins and negative control of cell growth. Activation of *p53* by DNA damage results in increased p21 levels. It can then bind to a number of cyclin–CDK complexes including cyclin D–CDK4, cyclin E–CDK2 and cyclin A–CDK2. This prevents phosphorylation of RB, keeping it in the activated state, and inhibits release of the transcription factor E2F which could otherwise promote transcription of genes involved in DNA synthesis (see above and Section 3.4.2). This causes the cell cycle to arrest in G_1. Clearly, loss or mutation of *p53* will lead to loss of this checkpoint control and cells will be able to enter S phase with damaged DNA [1,3].

4.4.3   S phase

Once cells enter S phase, a further set of cyclins and CDKs are required for continued DNA replication. In mammalian cells, cyclin A–CDK2 performs this function. Cyclin A is expressed from S phase through G_2 and M. Cyclin A binds to two different CDKs. Initially during S phase it is found complexed to CDK2 and is subsequently found during G_2 and M complexed to CDC2. In addition, it has been found in a multimeric protein complex together with the E2F transcription factor and the RB-related proteins p107 and p130.

Cyclin A has a role in both transcriptional regulation and replication. Cyclin A binds to the E2F transcription factor. Through this interaction, CDK2 phosphorylates the associated DP1 subunit and inhibits E2F's DNA binding activity. As cyclin A is synthesized during S phase it is therefore capable of down-regulating transcription of the genes activated at earlier stages of the cell cycle by E2F [11]. In addition, cyclin A has been shown to associate with the adenovirus E1A protein, a protein which can immortalize fibroblasts and, in cooperation with *RAS*, can lead to a transformed phenotype. This result has suggested that cyclin A might be involved in adenovirus-induced cell proliferation [11].

Cyclin A was one of the first cyclins to be implicated in tumor development. The cyclin A gene was found to be the unique insertion site for the hepatitis B virus in one clonal tumor [11] (see Section 7.4). The virus was found to integrate into the second intron of the gene resulting in the production of a chimeric protein in which the region, N terminal to the cyclin box, was replaced with viral sequences. This led to the removal of the 'destruction' box necessary for the degradation of the cyclin in mitosis. Gene rearrangements involving cyclin A have however rarely been observed in other liver tumors and there have been few reports of other abnormalities of cyclin A in tumors from other sites.

4.4.4   Mitosis

Entry into the final phase of the cell cycle, mitosis, is signaled by the activation of the cyclin B–CDC2 complex. This complex accumulates during S and G_2 but is kept in the inactive state by phosphorylation of tyrosine 15 and threonine 14 residues, a process regulated by the wee1/mik1-related kinases. At the end of G_2, the CDC25c phosphatase is stimulated to dephosphorylate these residues thereby activating CDC2 [4]. Cyclin B is located in the cytoplasm during interphase but is translocated to the nucleus at the beginning of mitosis. Cyclin B–CDC2 plays a major role in controlling the rearrangement of the microtubules in mitosis. In addition, the complex plays a role in disassembling the nucleus and allowing the cell to round up and divide.

Tumor cells can enter mitosis with damaged DNA, suggesting a defect in the G_2/M checkpoint. Tumor cell lines have been shown to activate the cyclin B–CDC2 complex irrespective of the state of the DNA and it has been suggested that a defect may lie in the inability of the tyrosine 15 and threonine 14 residues to be dephosphorylated [4].

There is one final checkpoint which occurs at the end of metaphase and at this point the correct assembly of the mitotic apparatus and the alignment of chromosomes on the metaphase plate are monitored. Normal cells arrest at this point if there are any defects, whereas in tumor cells abnormalities of spindle formation are found, suggesting that the checkpoint control is lost. Cyclin B is degraded as cells enter into anaphase. Re-establishment of interphase can then be initiated.

Rapid advances in the last few years have begun to show how disruption of cell cycle processes are involved in tumorigenesis. It is clear that many steps still remain to be identified to completely clarify the picture, particularly in the latter stages of the cycle. These will undoubtedly fall into place in the next few years. These areas may provide alternative targets for cancer therapy in the near future.

4.5   Mismatch repair genes

The importance of mismatch repair genes has been highlighted by the discovery that both sporadic and hereditary colorectal cancers show defects in these genes [12]. Hereditary nonpolyposis colon cancer (HNPCC) accounts for

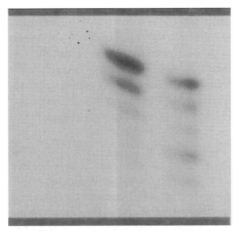

Figure. 4.4:   Microsatellite instability in colorectal cancer. One allele is seen in the DNA of normal tissue when tested with the microsatellite marker D2S123. In tumor DNA in the right hand track, additional bands are seen indicating the presence of replication errors. Figure courtesy of Dr Cecilia Brassett, Molecular Genetics Laboratory, Addenbrookes Hospital, Cambridge.

around 10–15% of all colorectal cancer (see Chapter 6). The first of four causative genes for this disease was mapped to chromosome 2p21 by linkage studies. A clue to the function of the gene which subsequently helped in its isolation came from studies of the tumors in these patients. When DNA from tumors was compared with DNA from normal tissues, the tumor DNA showed widespread alterations in short repeated sequences distributed throughout the genome [13]. These were seen as additional bands over and above the usual one or two alleles identified in the normal tissue DNA (*Figure 4.4*). This finding suggested that replication errors, caused by slippage of DNA polymerase, had occured during tumor development and had not been repaired. This phenotype was termed replication error positive (RER positive). Similarly, RERs were seen in sporadic colorectal cancers as well as other tumors such as endometrial, breast, prostate, lung and stomach [14,15,16]. The mechanism underlying this observation was suggested by previous studies of bacteria and yeast in which it had been shown that defects in mismatch repair genes resulted in instability in short repeated sequences.

   The best studied mismatch repair system is the DNA adenosine methylase (DAM)-instructed Mut HLS pathway in *Escherichia coli* (*Table 4.3*) [12]. DNA polymerase misincorporation errors occur during DNA replication and will result in the introduction of mismatches approximately one in every million nucleotides incorporated into DNA. These are normally repaired post-replication. Mismatch repair in *E. coli* occurs via excision of the newly synthesized strand containing the mismatch (*Figure 4.5*) The newly synthesized strand is identified by the lack of methylation (methylation of this strand by the DAM methylase normally occurs post-replication). The mis-

Table 4.3: Mismatch repair genes in bacteria, yeast and humans

Bacteria	Yeast	Human
MutS	*MSH2*	*hMSH2*
		GTBP
MutL	*MLH1*	*hMLH1*
MutL	*PMS1*	*hPMS1*
MutL	*PMS2*	*hPMS2*

Only the bacterial and yeast genes homologous to the human genes involved in cancer are shown. Other genes are involved in mismatch repair but human homologs have not so far been shown to cause tumors in man.

match is initially recognized by the MutS protein and is stabilized by the subsequent binding of MutL. MutH is then activated by this binding and the protein, which has single-stranded nuclease activity, nicks the unmethylated strand of DNA. The nicking is then followed by the excision step involving the ATP-dependent helicase UvrD and one of the single-stranded exonucleases (*exo*I, *exo*VII or *rec*J) to remove several thousand nucleotides. The DNA is then resynthesized by the polymerase III holoenzyme and finally the DNA nick is sealed by DNA ligase.

Saccharomyces cerevisiae has a similar mismatch repair system to that found in *E. coli* (*Table 4.3*) Two *MutS* homologs have been identified, *MSH1* and *MSH2* which function in the mitochondria and the nucleus respectively. Mutations in these genes lead to a high spontaneous mutation rate. There are several *MutL* homologs in yeast, notably *MLH1* and *MLH2* and *PMS1* and *PMS2*. *MLH1* acts in conjunction with *MSH2* and *PMS1* in the major mismatch repair pathway. In *S. cerevisiae*, mutations in the mismatch repair genes lead to destabilization of repetitive DNA of up to 700-fold [17]. This finding was a preliminary clue to suggest that the RER phenotype might be caused by defects in mismatch repair and eventually led to the identification of the genes involved in HNPCC.

Five mismatch repair genes have been identified in humans (*Table 4.4*). *hMSH2* lies at 2p21 (originally described as 2p15–16), the location initially identified in HNPCC families by linkage analysis [18]. Linkage analysis suggested the position of a second HNPCC gene at 3p21 and this was cloned by a degenerate PCR approach [19]. Two further human homologs of *MutL*, *hPMS1* and *hPMS2*, were then identified by the same method and mapped to chromosomes 2q and 7q respectively [20]. Most recently a fifth human mismatch repair gene, *GTBP*, which forms dimers with *hMSH2*, has been identified, although mutations in it have not yet been identified in colorectal tumors [21].

As already described in Chapter 3, the tumor suppressor genes are associated with loss of heterozygosity leading to defects in both copies of the gene. Initially no evidence of LOH was found at either 2p21 or 3p21, suggesting that the mismatch repair genes operated by a different mechanism. However it is

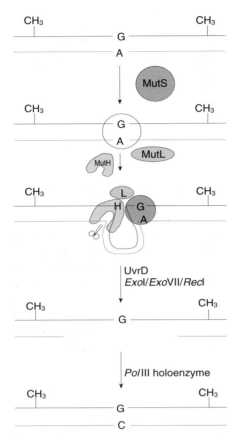

Figure. 4.5:   Mismatch repair in *E. coli*. The mismatch on the newly replicated strand (shown in orange) is recognized initially by the MutS protein. MutL subsequently binds to stabilize the structure and MutH cuts the unmethylated stand. The uvrD protein together with one of the single-stranded exonucleases removes 1–2 kb. The new strand is resynthesized by the *pol* III holoenzyme.

now clear that these genes also follow a version of the two-hit hypothesis leading to cells which are deficient in mismatch repair. Evidence for this came from a number of studies. Firstly, analysis of a number of tumors from individuals with germ-line mutations in either *hMSH2*, *hMLH1* or *hPMS1* demonstrated somatic mutations in the second copy of the gene [20,22,23]. Secondly, cell lines with a mutation in only one copy of *hMSH2* were found to be proficient in mismatch repair whereas those which were hemizygous for a mutation were repair deficient [24,25]. Finally loss of the wild-type allele has been shown to occur in 25% of tumors from patients with germ-line mutations in *hMLH1* [26].

Table 4.4: Human mismatch repair genes involved in HNPCC

Gene	Location	Size (base pairs)	Protein size (amino acids)
hMSH2	2p21	2727	909
hMLH1	3p21	2268	756
hPMS1	2q31–33	2795	932
hPMS2	7p22	2586	862

These studies all suggest that inactivation of both alleles of the mismatch repair genes are necessary for tumorigenesis.

Two studies have indicated how defects in these genes might lead to the development of tumors by showing that loss of these genes leads to a mutator phenotype. In other words, mutations in mismatch repair genes result in a higher than normal mutation rate, allowing the accumulation of mutations in other genes such as *p53* or *APC* [27,28]. These studies show once again how multiple genes are involved in the development of tumors and integrates the action of the mismatch repair genes with the oncogenes and tumor suppressor genes.

The use of the oncogenes, tumor suppressor genes, cell cycle regulators and mismatch repair genes in the diagnosis and in determining the prognosis of some of the commonest cancers is discussed in the following chapters.

References

1. Pines, J. (1995) *Adv. Cancer Res.*, **66**, 181.
2. Reed, S.I., Wittenberg, C., Lew, D.J., Dulic, V. and Henze, M. (1991) *Cold Spring Harbor Symp. Quant. Biol.*, **56**, 61.
3. Harper, J.W. and Elledge, S.J. (1996) *Curr. Opin. Genet. Dev.*, **6**, 56.
4. Hunter, T. and Pines, J. (1994) *Cell*, **79**, 573.
5. Hinds, P.W., Dowdy, S.F., Eaton, E.E., Arnold, A. and Weinberg, R.A. (1994) *Proc. Natl Acad. Sci. USA*, **91**, 709.
6. Leach, F.S., Elledge, S.J., Scherr, C.J., Wilson, J.K., Markowitz, S., Kinzler, K.W. and Vogelstein, B. (1993) *Cancer Res.*, **53**, 1986.
7. Khatib, Z.A., Matsushime, H., Valentine, M., Shapiro, D.N., Sherr, C.J. and Look, A.T. (1993) *Cancer Res.*, **53**, 5535.
8. Serrano, M., Hannon, G.J. and Beach, D. (1993) *Nature*, **366**, 704.
9. Hannon, G.J. and Beach, D. (1994) *Nature*, **371**, 257.
10. Sherr, C.J. (1994) *Cell*, **79**, 551.
11. Brechot, C. (1993) *Curr. Opin. Genet. Dev.*, **3**, 11.
12. Fishel, R. and Kolodner, R.D. (1995) *Curr. Opin. Genet Dev.*, **5**, 382.
13. Aaltonen, L.A., Peltomaki, P., Leach, F.S., Sistonen, P., Pylkkanen, S.M., Mecklin, J.P. (1993) *Science*, **260**, 812.
14. Ionov, Y., Peinado, M.A., Malkbosyan, S., Shibata, D. and Perucho, M. (1993) *Nature*, **363**, 558.
15. Thibodeau, S.N., Bren, G. and Schaid, D. (1993) *Science*, **260**, 816.

16. Wooster, R., Clenton-Jansen, A.M., Collins, N. *et al.* (1994) *Nature Genetics*, **6**, 152.
17. Strand, M., Prolla, T.A., Liskay, R.M. and Petes, T.D. (1993) *Nature*, **365**, 274.
18. Fishel, R., Lescoe, M.K., Rao, M.R.S., *et al.* (1993) *Cell*, **75**, 1027.
19. Papadopoulos, N., Nicolaides, N.C., Wei, Y.-F., Ruben, S.M., Carter, K.C. and Rosen, W.A. (1994) *Science*, **263**, 1625.
20. Nicolaides, N.C., Papadopoulos, N., Liu, B., *et al.* (1994) *Nature*, **371**, 75.
21. Palombo, F., Gallinari, P., Iaccarino, I. *et al.* (1995) *Science*, **268**, 1912.
22. Leach, F.S., Nicolaides, N.C., Papadopoulos, N. *et al.* (1993) *Cell*, **75**, 1215.
23. Liu, B., Parsons, R.E., Hamilton, S.R. *et al.* (1994) *Cancer Res.*, **54**, 4590.
24. Parsons, R., Li, G.-M., Longley, M.J. *et al.* (1993) *Cell*, **75**, 1227.
25. Umar, A., Boyer, J.C., Thomas, D.C. *et al.* (1994) *J. Biol. Chem.*, **269**, 14367.
26. Hemminki, A., Peltomaki, P., Mecklin, J.-P. *et al.* (1994) *Nature Genetics*, **8**, 404.
27. Bhattacharya, N.P., Skandalis, A., Ganesh, A., Groden, J. and Meuth, M. (1994) *Proc. Natl Acad. Sci. USA*, **91**, 6319.
28. Lazar, V., Grandjouan, S., Bognel, C. *et al.* (1994) *Hum. Mol. Genet.*, **3**, 2257.

Chapter 5

Lung cancer

5.1 Introduction

Lung cancer is one of the most prevalent cancers in the developed nations of the world and it is a leading cause of cancer deaths. The age-adjusted death rates for the major industrialized countries are shown in *Table 5.1*.

Table 5.1: Age-adjusted death rates in the major industrialized countries of the world

Country	Age-adjusted death rates (1990–1993) per 100 000 population	
	Male	Female
Italy[b]	57.0	7.3
Canada[a]	55.1	21.8
USA[a]	57.1	25.6
UK[c]	55.9	21.0
Germany	47.9	8.4
France[a]	47.0	5.2
Japan	30.6	8.1

Abstracted from ref. 1.
[a] 1990–1992; [b] 1990–1991; [c] 1992–1993.

Statistics from the USA illustrate the problem. In 1996 an estimated 177 000 new cases of lung cancer will be diagnosed and the number of deaths attributable to this disease are estimated to be 158 700 [1]. From the 1950s to the 1980s, American males experienced a 184% increase in age-adjusted lung cancer deaths and females showed a 360% increase over the same period, surpassing the breast cancer death rate for the first time in over 50 years [2]. This trend is also seen in the lung cancer mortality rate for many other countries [3].

There are many different types of malignant lung tumors and the World Health Organisation (WHO) classification, recommended originally in 1977, is usually accepted as the definitive classification of this cancer [4]. As with colorectal and gastric cancers, there has been little improvement in survival rates over the last 35 years.

Non-small-cell lung cancer (NSCLC) was the term adopted by the WHO in 1977 to group together adenocarcinoma, squamous cell carcinoma and large cell carcinoma. NSCLC accounts for approximately 75–80% of all lung cancers. The prognosis for patients with NSCLC is poor, with an overall 5 year survival of approximately 15% [5]. Small-cell lung cancer (SCLC) comprises the remaining 20–25% of lung cancer cases and has been recognized as a distinct clinicopathological entity for over 20 years. It differs markedly from the other histological subtypes of lung cancer, both in its biological behavior and in its responsiveness to chemotherapy and radiotherapy. More than any other type of lung cancer, SCLC is associated with cigarette smoking. Its natural history is characterized by a very aggressive clinical course and a propensity for early widespread dissemination and for this reason it is regarded as a 'systemic' disease at diagnosis [6]. The median survival of untreated patients ranges from 6 to 17 weeks and less than 0.5% of patients survive for 5 years following surgical treatment alone. Although very sensitive to radiotherapy and combination chemotherapy the vast majority of patients relapse and die of their disease. Cure therefore remains an elusive goal for most patients and new therapeutic strategies and directions are urgently needed [6].

5.2   Genetic changes in lung cancer

Many cytogenetic abnormalities occur in lung cancer, one of the most frequently recognized changes being a deletion on the short arm of chromosome 3 (3p14–p23) which is found in approximately 90% of SCLC and 50% of NSCLC cases [7]. The 3p14–p23 deletion appears to be an important early event in the development of lung cancer and the size of the area lost by chromosomal deletion or translocation is large (many megabases) suggesting that more than one candidate tumor suppressor gene may be involved [6, 8 and references therein]. Deletions on 3p and 9p (see Section 5.2.4) are presently the primary target for positional cloning of putative tumor suppressor genes. A recent study in lung cancer cell lines has confirmed that 11p harbors several putative tumor suppressor genes which become inactivated at different stages of tumor development [9]. Allele loss on chromosome 3 has also recently been demonstrated to precede damage to the *p53* gene [10].

Allelic deletion mapping has implicated the involvement of at least two other tumor suppressor genes on chromosome 5 in lung tumorigenesis, including the *APC* gene at 5q21 [11].

5.2.1   RB1 *and* p53

Patients with familial retinoblastoma develop second tumors including lung cancer and it has been found that there is a 15-fold increased risk of developing lung cancer, particularly SCLC, in familial retinoblastoma patients cured of their retinoblastoma [12]. These *RB1* mutations thus represent the first identified inherited abnormalities predisposing to lung cancer [8].

Regions showing loss of heterozygosity (LOH) by RFLP analysis include 11p and 13q (*RB1*), and 17p (*p53*). The *RB1* gene and its product are altered in nearly all cases of SCLC and approximately 10–30% of cases of NSCLC while the *p53* gene is mutated in almost 100% of SCLCs and approximately 50% of NSCLC cases [6, 8, 13 and references therein].

A high proportion of NSCLCs express p53 and BCL2 proteins and there is an inverse relationship between their expression. Although no correlation has been found between p53 expression and overall survival, p53 protein accumulation occurs in tumors with metastatic nodal involvement or in patients who develop metastases during follow-up [14]. The results of this study indicated that the probability of survival was higher in patients with NSCLC who expressed the BCL2 protein. However, in a study in SCLC, the BCL2 protein was found to be present in most cases and to be, therefore, of little value as a prognostic or treatment marker [15].

A study of 247 surgically treated patients with NSCLC, showed that *p53* overexpression was associated with high tumor grade and the presence of lymph node metastasis, but not with advanced tumor stage. Survival was no different between p53-positive and p53-negative tumors within node-positive tumors. In contrast, survival was significantly better in p53-negative tumors than in p53-positive tumors within the group of node-negative tumors. These authors concluded, therefore, that p53 overexpression is an independent prognostic factor in node-negative NSCLC [16].

In addition, CYP1A1 (human cytochrome p450) germ-line polymorphisms, which have previously been shown to be associated with a genetic predisposition to lung cancer (see Section 5.3), have recently been demonstrated also to be related to cigarette smoking-associated *p53* mutations [17].

5.2.2 MYC

Common dominant oncogene mutations in lung cancer include mutations in *RAS* and *MYC* family members. Abnormal expression of all three members of the *MYC* family (*NMYC*, *LMYC* and *MYC*) has been detected in lung cancers. Amplification of one or other of the *MYC* genes is found in 10–20% of lung tumors. The role of *MYC* overexpression in lung cancer requires further definition, however, the current evidence suggests that *MYC* is involved in the progression of SCLC [18]. Amplification of *MYC* is seen frequently in SCLC but it is an uncommon event in NSCLC, the general mechanism of activation being amplification with resultant overexpression [19].

There have been several reports from Japan of RFLP analysis of *LMYC* as a marker for metastatic potential in NSCLC. Poor prognosis was related to the type of RFLP pattern, L-L, homozygous for the 10 kb *LMYC* fragment; S-S, homozygous for the 6 kb *LMYC* fragment, or L-S, heterozygous for the *LMYC*

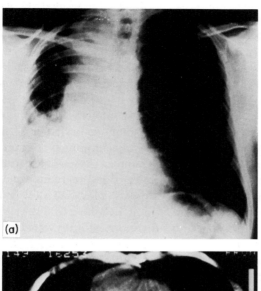

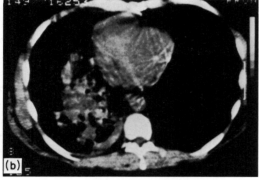

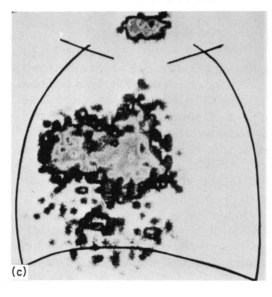

Figure 5.1: (a) X-ray showing primary tumor in right lung and mediastinal metastases, (b) CT scan, and (c) image of lung tumor following localization of ^{131}I-labeled antibody to p62. Reproduced from ref. 21 with permission from Stockton Press.

fragment, with those adenocarcinomas with the L-S or S-S pattern having the highest incidence of lymph node metastasis and metastasis to other organs [19].

The tumors of SCLC patients with amplification of *MYC* have been shown to be the more aggressive and these patients have consequently had a significantly worse prognosis [20]. Amplification and overexpression of *MYC* in SCLC is more common in previously treated patients than it is at initial presentation, indicating that it is a relatively late event in pathogenesis, perhaps contributing to the more aggressive behavior observed at relapse [6].

Increased *NMYC* expression in SCLC detected by *in situ* hybridization has been correlated with poor response to chemotherapy, rapid tumor growth and short survival times [20]. Elevated *MYC* expression has also been demonstrated in 19% of SCLCs and 42% of NSCLCs using the Myc1-9E10 antibody, but no correlation between elevated expression and survival could be shown in these patients [20].

As survival from metastatic lung cancer is so poor, enhancers of chemotherapy have been sought. As a preliminary to targeted therapy with antibodies, the ability of monoclonal antibody Myc1-6E10 to localize lung tumors was investigated. Following radiolabeling with ^{131}I, this antibody was administered intravenously to 20 patients with malignant disease. γ-Camera imaging 24 and 48 h later showed good tumor localization in 12 out of 14 patients who had primary bronchial carcinoma (*Figure 5.1*). In the remaining six patients who had pulmonary metastases from a variety of tumors, no localization was found. As the p62 protein is located primarily in the cell nucleus, the localization was presumed to depend on release of nuclear contents following cell death [21]. Although this protein is not the ideal target for such studies, it was the first to show the feasibility of using oncogenes or their products as targets for this approach (see Chapter 12).

5.2.3 RAS

Mutations in *RAS* family members, particularly *KRAS* codon 12 mutations, are found in over 30% of NSCLCs but do not normally occur in SCLCs [6, 8 and references therein]. Patients with SCLCs that are found to have a *RAS* mutation have a significantly poorer survival than patients lacking a *RAS* mutation [22]. Increased expression of the *RAS* gene has been detected in lung cancers, as have structural abnormalities in the gene. Differential expression of *RAS* p21 determined by Western blotting with Y13 259 has been correlated with histological classification. In one study, 82% of squamous cell cancers had increased levels of p21 whereas only 8% of nonsquamous cell tumors showed any increase in expression, thus correlating increased expression with histological classification [23].

A number of studies have indicated that the presence of a *KRAS* mutation, or enhanced expression of the oncogene product, is an independent, adverse prognostic factor for survival [19, 24 and references therein]. PCR has been used to identify point mutations in *KRAS* in NSCLCs, primarily in adenocarcinomas, and has shown that carcinogens in tobacco smoke were the cause of the mutations as there was a strong correlation between the incidence of mutations and the smoking habits of patients. However, in contrast to the studies cited above, primary NSCLCs with *KRAS* mutations were shown to be smaller and to have a lower metastatic potential than tumors without *RAS* mutations. If the presence of *KRAS* mutations can be confirmed as a good prognostic indicator it may be of use in planning chemotherapy for certain patients [20].

RFLPs associated with *HRAS* have been shown to be possible genetic markers for lung cancer. Four common alleles are identified in the population along with a number of rarer alleles. An abnormal allele distribution has been associated with the more aggressive NSCLCs compared with normal cells or SCLCs, suggesting a degree of genetic predisposition to this disease [25]. A similar RFLP has been associated with *LMYC* and one particular allele has been associated with metastatic disease.

5.2.4   *Cyclin-dependent kinase 4 inhibitor (CDK4I)*

Deletions and translocations of chromosomal region 9p21 have been associated with lung cancer and chromosome 9p loss now represents the most common genetic change in NSCLC [26]. Subsequent analysis of the deleted regions has identified a loss that contains the human cyclin-dependent kinase 4 inhibitor (*CDK4I*), also known as *p16*, *MTS1* and *CDKN2*, which is involved in the G_1/S transition of the cell cycle [26] (Chapter 4).

Homozygous deletion of *CDK4I* has been detected in 23% of cell lines established from patients with NSCLC compared with 1% of cell lines established from patients with SCLC [13]. In addition, this homozygous deletion was observed only in cell lines from patients with stage III or stage IV NSCLC and coincident loss of *CDK4I* genes and functional RB protein was only seen in two out of 135 cell lines examined. The frequency of homozygous *CDK4I* gene deletion in NSCLC cell lines is greater than that observed for any known, or candidate, tumor suppressor gene [13].

From comparisons with previously published data on other chromosomal abnormalities in preneoplastic and neoplastic tissue from NSCLCs, it has been shown that LOH at 3p and 9p loci occurs early in the hyperplasia stage, but *RAS* gene point mutations occur relatively late, at the carcinoma *in situ* stage of NSCLC [27]. These authors concluded that LOH at 9p occurs at the earliest stage in the pathogenesis of lung cancer and involves all regions of the respiratory tract. LOH in NSCLC is not random but targets a specific allele in individuals. Studying preneoplastic lesions may help to identify intermediate markers for risk assessment and chemoprevention [27]. Thus, strategies aimed at the detection of rare neoplastic cells in sputum samples, harboring absent p16 expression secondary to methylation or homozygous deletion of the gene, might be used for the early detection of lung cancer [26].

5.2.5 HER2/NEU/ERBB2

The *HER2/NEU* oncogene (also called *ERBB2*) encodes the transmembrane protein p185^{neu} which shares extensive homology with the epidermal growth factor receptor (EGFR). High levels of EGFRs have been associated with NSCLCs, particularly squamous carcinomas. The strongest staining seen with an anti-EGFR antibody was associated with stage III tumors, suggesting, as with other tumor types, that expression of the EGFR may be associated with growth and metastasis. An antibody to the external domain of the receptor may prove to be better for targeting purposes than the antibody to p62 for the localization of lung tumors. High concentrations of the EGFR in these tumors with poor prognosis also suggests a possible target for therapy [28].

NEU expression occurs in approximately one-third of NSCLCs and, in adenocarcinomas, it has been shown to be an independent unfavorable prognostic factor [18, 19, 24, 29 and references therein].

5.3 Other potential markers in lung cancer

Cancer risk from exposure to tobacco smoke varies widely from person to person, depending in part on the status of particular genes and acquired susceptibilities. Certain genes determine how cells activate and detoxify carcinogens. Activated carcinogen metabolites may bind to DNA and form DNA adducts many of which can induce genetic mutations. Recent studies have indicated that the levels of two different carcinogen-DNA adducts vary in lung tissue in association with three separate genetic polymorphisms (*CYP2D6, CYP2E1, GSTMI*). *CYP2D6* and *CYP2E1* are cytochrome p450 genes and *GSTMI* is the gene for glutathione S-transferase. The results suggest that the genetic polymorphisms are predictive of carcinogen–DNA adduct levels and would therefore be predictive of an individual's lifetime response to carcinogen exposure [30].

Finally, telomerase activity may be useful as a diagnostic marker for detecting the existence of immortal lung cancer cells and as a target for therapeutic intervention in the future [31].

A number of recent reviews cover molecular events [18], the biology of SCLC [32], oncogenic alterations [33] and prognostic factors [19, 24, 29] in lung cancer and the reader is referred to them for more details.

References

1. Parker, S.L., Tong, T., Bolden, S.B. and Wingo, P.A. (1996) CA – *Cancer J. Clinic.*, **46**, 5.
2. Perera, F.F., Santella, R., Brandt-Rauf, P., Kahn, S., Jiang, W. and Mayer, J. (1991) in *Origins of Human Cancer: a Comprehensive Review.* Cold Spring Harbor Laboratory Press, Cold Spring Harbor, NY, p. 219.
3. Gilliland, F. and Samet, J.M. (1994) in *Cancer Surveys,* Vol. 19: *Trends in Cancer Incidence and Mortality* (R. Doll, J.F. Fraumeni, Jr, and C.S. Muir, Eds). Cold

Spring Harbor Laboratory Press, Cold Spring Harbor, NY, p. 175.
4. Minna, J.D., Pass, H., Glatstein, E. and Ihde, D.C. (1989) in *Cancer. Principles and Practice of Oncology,* 3rd Edn (V.T. DeVita, S. Hellman, and S.A. Rosenberg, Eds). Lippincott, Philadelphia, p. 591.
5. Jahangiri, M., Barton, R., and Goldstrow, P. (1995) in *Oncology. A Multi-disciplinary Textbook* (A. Horwich, Ed.). Chapman & Hall Medical, London, p. 599.
6. Ellis, P.A. and Smith, I.E. (1995) in *Oncology. A Multidisciplinary Textbook.* (A. Horwich, Ed.). Chapman & Hall Medical, London, p. 585.
7. Whang-Peng, J., Knutsen, T., Gazdar, A., Steinberg, S., Oie, H., Linnoila, I., Mulshine, J., Nau, M. and Minna, J. (1991) *Genes Chrom. Cancer,* **3**, 168.
8. Minna, J., Maneckjee, R., D'Amico, *et al.* (1991) in *Origins of Human Cancer: a Comprehensive Review.* Cold Spring Harbor Laboratory Press, Cold Spring Harbor, NY, p. 781.
9. Bepler, G. and Koehler, A. (1995) *Cancer Genet. Cytogenet.,* **84**, 39.
10. Chung, G.T.Y., Sundaresan, V., Hasleton, P., Rudd, R., Taylor, R. and Rabbitts, P.H. (1995) *Oncogene,* **11**, 2591.
11. Wieland, I., Bohm, M., Arden, K.C., Ammermuller, T., Bogatz, S., Viars, C.S. and Rajewsky, M.F. (1996) *Oncogene,* **12**, 97.
12. Sanders, B., Jay, M., Draper, G. and Roberts, E. (1989) *Br. J. Cancer,* **60**, 358.
13. Kelley, M.J., Nakagawa, N., Steinberg, S.M., Mulshine, J.L., Kamb, A. and Johnson, B.E. (1995). *J Natl Cancer Inst.,* **87**, 756.
14. Fontanini, G., Vignati, S., Bigini, D., Mussi, A., Lucchi, M., Angeletti, C.A. and Bevilacqua, G. (1995) *Br. J. Cancer,* **71**, 1003.
15. Yan, J.-J., Chen, F.-F., Tsai, Y.-C. and Jin, Y.-T. (1996) *Oncology,* **53**, 6.
16. Dalquen, P., Sauter, G., Torhorst, J. *et al.* (1996) *J. Pathol.,* **178**, 53.
17. Kawajiri, K., Eguchi, H., Nakachi, K., Sekiya, T. and Yamamoto, M. (1996) *Cancer Res.,* **56**, 72.
18. Roth, J.A. (1995) *Lung Cancer,* **12** (Suppl. 2), S3.
19. Mountain, C.F. (1995) *Chest,* **108**, 246.
20. Viallet, J. and Minna, J.D. (1990) *Am. J. Res. Cell Molec. Biol.,* **2**, 225.
21. Chan, S.Y.T., Evan, G.I., Titson, A., Watson, J., Wraight, P. and Sikora, K. (1986) *Br. J. Cancer,* **54**, 761.
22. Slebos, R., Kibbelaar, R., Dalesio, O. *et al.* (1990) *N. Engl. J. Med.,* **323**, 561.
23. Field, J.K. and Spandidos, D.A. (1990) *Anticancer Res.,* **10**, 1.
24. Kanters, S.D.J.M., Lammers, J.-W. and Voest, E.E. (1995) *Eur. Respir. J.,* **8**, 1389.
25. Heighway, J., Thatcher, N., Cerny, T. and Hasleton, P.S. (1986) *Br. J. Cancer,* **53**, 453.
26. Sidransky, D. (1995) *J Natl Cancer Inst.,* **87**, 1201.
27. Kishimoto, Y., Sugio, K., Hung, J.Y., Virmani, A.K., McIntire, D.D., Minna, J.D. and Gazdar, A.F. (1995) *J Natl Cancer Inst.,* **87**, 1224.
28. Veale, D., Ashcroft, T., Marsh, C., Gibson, G.J. and Harris, A.L. (1987) *Br. J. Cancer,* **55**, 453.
29. Scagliotti, G.V., Masiero, P. and Pozzi, E. (1995) *Lung Cancer,* **12** (Suppl. 1), S13.
30. Kato, S., Bowman, E.D., Harrington, A., Blomeke, B. and Shields, P.G. (1995) *J. Natl Cancer Inst.,* **87**, 902.
31. Hiyama, K., Hiyama, E., Ishioka, S., Yamakido, M., Inai, K., Gazdar, A.F., Piatyszek, M.A. and Shay, J.W. (1995) *J. Natl Cancer Inst.,* **87**, 895.
32. Johnson, B.E. and Kelley, M.J. (1995) *Lung Cancer,* **12** (Suppl. 3), S5.
33. Strebhardt, K., Holtrich, U., Brauninger, A., Karn, T., Bohme, B., Doermer, A. and Rubsamen-Waigmann, H. (1994) *Oncol. Rep.,* **1**, 195.

Chapter 6

Colorectal cancer

6.1 Introduction

Colorectal cancer is the most common cause of death from cancer in the non-smoking male and female populations in the Western world. The age-adjusted death rates in the major industrialized countries of the world are shown in *Table 6.1* [1]. There has been little change in death rates over the last 50 years because, although there have been some improvements in survival, these have been masked by the increase in the incidence of the disease. The 5-year survival for colon cancer remains around 40% because of the late detection of the disease. Molecular techniques are beginning to help in the understanding of the disorder and are perhaps more advanced than for any other cancer. There are two reasons for this. Firstly it has been recognized for over a quarter of a century that colorectal tumors develop from adenomas, the so-called adenoma-carcinoma sequence [2]. This has meant that molecular changes can be more clearly delineated than in other cancers. Secondly, there are a number of inherited forms of colorectal cancer, the genes for which have now been isolated. Both oncogenes and tumor suppressor genes therefore play a role in colorectal cancer as discussed below. However it must be remembered that environmental and dietary factors also play a major role in this disease [3].

The disease is classified according to the Dukes' classification into Dukes' A–D cancers. The use of any of the oncogenes or tumor suppressor genes as prognostic indicators therefore has to be compared with this classification which remains the most useful method to date of determining survival.

Table 6.1: Age-adjusted death rates in the major industrialized countries of the world

Country	Age-adjusted death rates (1990–1993) per 100 000 population	
	Male	Female
Germany	21.3	15.1
UK[c]	20.3	13.6
France[b]	17.4	10.1
Canada[b]	16.9	11.2
USA[b]	16.5	11.2
Japan	15.7	9.8
Italy[a]	15.3	9.9

[a]1990–1991; [b]1990–1992; [c]1992–1993.

6.2   Genetic changes in sporadic colorectal cancer

6.2.1   MYC

Early studies of colorectal cancer using cell lines revealed amplification of *MYC* and *MYB* but the lines used probably represented atypical cases. Subsequently, no major role has been identified for either amplification or rearrangement of any oncogene in colorectal cancer [4,5].

Increased levels of *MYC* transcripts have been detected by Northern blotting in up to 70% of colorectal cancers, but have not correlated with disease progression, histological diagnosis or Dukes' staging [4,5]. A statistically significant association was found between the presence of synchronous polyps and overexpression of *MYC* although the implications of this are not clear [5].

Immunohistochemical studies with the monoclonal antibodies Myc1-6E10 and Myc1-9E10 have confirmed the Northern blotting results by showing high levels of the MYC protein, p62, in colorectal cancers (*Figure 6.1*). Some studies have found a correlation with tumor differentiation but others found no difference. In addition, increased levels of p62 have also been associated with inflammatory conditions such as Crohn's disease and ulcerative colitis. Some uncertainty has arisen in the literature concerning the use of Myc1-6E10 because it appears to recognize a protein present in both the nucleus and the cytoplasm. Suggestions for this unexpected distribution of p62 have included poor tissue fixation conditions or extraction of p62 by salt and/or low temperatures. Alternatively, it is possible that the antibody cross-reacts with an unre-

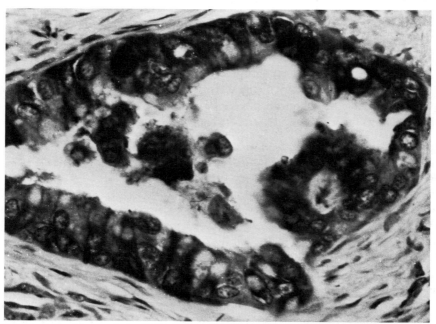

Figure 6.1:   Expression of MYC p62 in colorectal cancer.

lated antigen. A review of the staining patterns seen with this antibody in colorectal cancer has suggested that the observed expression is largely a function of the fixation process; the protein recognized is probably p62, but the possibility of nonspecific binding has not been totally excluded [4].

Studies of the RNA and protein levels confirm that a high level of *MYC* is a marker of cell proliferation. Whether deregulation and high expression of *MYC* are required to maintain the tumorigenic state has yet to be determined. With a few exceptions, there have been very few diagnostic or prognostic implications of *MYC* expression levels in colorectal cancers.

Two studies have suggested a role for *MYC* as a marker for monitoring the transition from the benign to malignant state. Levels of *MYC* transcripts determined by Northern blotting were closely related to histological type and size of colorectal polyps. In particular, adenomas containing carcinoma *in situ* or high-grade dysplasia expressed high levels of *MYC* [6,7].

Elevated expression of the *MYC* gene has also been particularly associated with tumors of the distal part of the colon. On the basis of this observation it has been suggested that elevated levels of *MYC* are a marker of a sub-group of sporadic colorectal cancers with a different etiology to those occurring on the right side of the colon. Additional studies following on from this observation have indicated loss of *MYC* regulation in carcinomas deleted for the chromosomal region containing the *APC* gene. These results are consistent with the possibility that the product of the *APC* gene exerts its effect by deregulating *MYC* [8]. This has been tested by the introduction of chromosome 5 by microcell fusion into colorectal cell lines which led to suppression of *MYC* deregulation [9].

6.2.2 RAS

Alterations in *RAS* have been studied extensively in colorectal cancer. There has been no evidence for amplification or rearrangement of the gene, but elevated expression of *RAS* and frequent mutations in both *KRAS* and *HRAS* have been identified.

One study demonstrated elevated levels of the *RAS* gene family transcripts in both premalignant and malignant tumors of the colon and rectum, suggesting that elevated expression might be critical in the process of carcinogenesis. As all polyps do not progress, despite showing increased *RAS* expression, the observation is consistent with the involvement of other genes in the development of cancers. In contrast, a second study found elevated *KRAS* and *HRAS* in only a small percentage of carcinomas and increased levels of *KRAS* in only one polyp. There was a correlation between poor prognosis and elevated levels of *KRAS* and *HRAS* if increased levels of *FOS* and *MYC* transcripts were also present in the same tumor. As both studies were small it is difficult to determine the significance of these observations [4].

Many immunohistochemical studies have been carried out either by Western blotting techniques or by immunohistochemistry. Although these studies have shown increased levels of RAS in tumors, this is most likely to be associated with cellular proliferation associated with carcinogenesis rather than the involvement of increased levels of RAS in colorectal cancer development.

Analysis of point mutations in the *RAS* gene has provided more consistent results and suggests that *RAS* mutations occur early in the development of colorectal tumors. The use of the PCR has allowed the selective amplification of specific sequences and has undoubtedly contributed to the reliability of the analysis. As small amounts of DNA can be analyzed it has been possible to select the areas of tissue to be used carefully, by extracting the DNA from tumor-rich regions of histological specimens without including areas of inflammatory or normal cells [10]. Using these techniques, 37–60% of colorectal tumors have been found to have an activated *RAS* gene, the majority of the mutations occurring in codons 12, 13 or 61 of *KRAS* [11].

Confirmation that mutations in *RAS* are an early event came from the finding that a high proportion of *RAS* mutations occur in adenomas, in tumors originating in adenomas and in both the benign and malignant areas of the same tumor. The mutational event appears to occur before the tumor becomes aneuploid [2]. A large study has investigated the timing of these events in colorectal cancer in relation to other genetic alterations. Approximately 50% of adenomas contain a *RAS* gene mutation and a similar mutation frequency is found in carcinomas but mutations have only been found in 9% of those adenomas of less than 1 cm in size [11]. This suggested the possibility that *RAS* mutations might be required for the conversion of small adenomas to larger ones by clonal expansion of the cell carrying the mutation. Subsequent studies have detected *RAS* mutations in the grossly normal mucosa of patients with colorectal cancer [12]. This was subsequently shown to be due to the presence of aberrant crypts which were present in the grossly normal appearing mucosa [13]. These aberrant crypts may be a precursor of a subset of colorectal tumors [13] but this remains to be proven. There is some similarity between these lesions and hyperplastic foci found in pancreatic cancer (see Section 7.3.1). In these lesions, *RAS* mutations have also been found but at too high a frequency to suggest a causative role in pancreatic cancer.

In a large study of over 100 patients no correlation was found between the presence of *RAS* mutations and the 5-year survival rate [14]. A further study showed evidence of *KRAS* mutations at codon 12 in 24% of cancers but again found no correlation with survival [15]. These authors did however find a decrease in survival if patients had both *RAS* and *p53* mutations. In a retrospective study of 247 primary and 166 metastatic colorectal cancers in which the regions containing the commonly mutated exons were sequenced, mutations were found in 37% of primary cancers and 20–59% of metastases. Although a number of different amino acid substitutions were found in the primary tumors the metastases were almost exclusively associated with glycine to aspartic acid alterations. The follow up in this study was only for 12 months, nonetheless, only two cases without a *KRAS* mutation developed metastases whereas 16 out of 29 with a mutation did [16]. Deaths predominantly occurred in the group with glycine to aspartic acid mutations. In contrast a second study found that G to A transitions (that is, changing the amino acid for glycine to aspartic acid) were associated exclusively with Dukes' B cancers whereas G to T and G to C changes were associated with metastatic behavior [17].

As *RAS* mutations are a relatively early event in the development of colorectal cancer (see Section 3.6.1), the possibility of detecting *RAS* mutations in stool samples as a screening method to detect cancers at an early stage has been investigated [18]. Preliminary data have shown that it was possible to detect *RAS* mutations in the stools of eight out of nine patients whose cancer had a known *RAS* mutation. Two of the positive cases had only adenomas present at the time of testing, suggesting that this technique could be a potential screening method, particularly if other genes involved in colorectal carcinogenesis could be included. A second study has confirmed the feasibility of this technique by identifying *RAS* mutations in the stools of approximately 50% of cases studied [19]. It remains to be seen how feasible this technique is as a screening tool when studies are extended to include larger numbers of cases and potentially to include other genes such as *APC* which is found mutated at an earlier stage than *RAS*. The nature of mutations found in *APC* in colorectal cancer (see Section 6.2.6) may preclude its use in screening unless novel methods for mutation detection can be developed.

6.2.3   p53

Loss of chromosome 17p is a common finding in colorectal cancer and occurs at a relatively late stage in tumor progression [20]. A number of studies have shown that LOH of 17p, presumed to include *p53*, is associated with poor prognosis [14,21] whilst others have found that it also correlates directly with the presence of metastases [22].

p53 expression in colorectal cancers has been studied extensively using immunohistochemical techniques, the assumption being that mutations in *p53* result in a stable protein with a prolonged half-life. The correlation between nuclear p53 expression and survival has however been contradictory. Whilst some groups have shown that increased expression is associated with shortened survival [23, 24] others have shown no correlation [15]. The explanation for this discrepancy is possibly due to variations in the antibodies used, the site of staining and more critically, may also depend on whether antibody staining is a good indicator of p53 mutations, a point which has been debated [25]. Different procedures and antibodies have been evaluated and antibody DO-7 shown to be the most sensitive and specific for immunohistochemical studies [26]. In a study comparing staining obtained with DO-7 versus a polyclonal antibody and by objectively analyzing staining by computerized image analysis, p53 expression was shown to have no prognostic significance [27]. Comparison of antibody staining using DO-7 and mutation analysis by single-stranded conformation polymorphism (SSCP) showed only a 69% correlation between the two techniques and again no correlation between p53 and survival was seen [28]. In addition, the association between increased p53 levels and poor prognosis in one study [29] was related to cytoplasmic staining. However, the majority of other studies have found no convincing evidence for p53 staining in the cytoplasm [27].

In light of these findings, because mutations in *p53* which lead to aberrant splicing and protein truncation do not alter intracellular levels of the p53 protein and because allelic loss at 17p can occur in the absence of *p53* mutations, several studies have looked instead at mutations in *p53* at the DNA level. The timing of such changes and any association with survival have then been correlated. Mutations in *p53* were found more frequently in tumors of the distal colon [30] again suggesting a difference in the etiology of tumors occuring in different regions of the colon (see Sections 6.2.1 and 6.2.7). A clear association between poor prognosis and the presence of *p53* mutations was also shown in this study, so direct mutation analysis may after all improve prognostic accuracy [30]. *p53* mutations appear to occur at the early colorectal cancer stage, that is in the conversion of adenomas into early cancers [31]. Analysis of *p53* mutations have shown that changes occur in colorectal cancers prior to aneuploid clonal divergence of the tumor [32].

6.2.4   MCC

The 'mutated in colorectal cancer' gene (*MCC*) was originally isolated in a search for the causative gene for familial adenomatous polyposis (FAP). Although it was shown not to be involved in the inherited condition, mutations in *MCC* have been identified in a proportion of sporadic colorectal cancers [33]. However studies have shown that this gene does not appear to have a major role in prognosis or diagnosis of colorectal cancer [34].

6.2.5   DCC

The long arm of chromosome 18 has been found to be deleted in over 70% of colorectal cancers. This led to the isolation in 1990 of the 'deleted in colorectal cancer' (*DCC*) gene [35]. The DCC protein shows features in common with the neural cell adhesion molecule and may therefore interact with other proteins involved in cell–cell and cell–matrix interactions. Loss or mutation of this gene could therefore result in impaired contacts between cells, contributing to tumor growth and invasion.

Immunohistochemical studies have shown that *DCC* is present in hyperplastic and most adenomatous polyps but absent in colorectal cancers. The majority of studies have shown that LOH on 18q is associated with decreased survival [36]. In particular, Dukes' B patients who show no allele loss at this position had a 5-year survival of 93%, a result similar to that expected for

Dukes' A patients. In contrast, LOH at 18q in Dukes' B patients resulted in a very poor survival similar to that for Dukes' C patients and it was suggested that this group could therefore benefit from adjuvant therapy [37]. In addition, expression of *DCC* has been shown to be absent in colorectal cancers metastatic to the liver but present in the majority of nonmetastatic tumors [38] so its absence may play a role in the development of distant metastases.

6.2.6 APC

Loss of heterozygosity of chromosome 5q has been found in around 50–60% of colorectal cancers as well as in adenomas [39], suggesting the involvement of this region in the development of sporadic colorectal cancer as well as in familial adenomatous polyposis (FAP) (see below). Following the cloning of the *APC* gene this has been confirmed by the identification of mutations in around 60% of sporadic colorectal cancers and adenomas [40]. In addition, the second copy of *APC* is either lost or can be inactivated by a second mutation. Although one study has shown that this occurs very early in the development of adenomas [41] most have suggested that the second 'hit' occurs with increasing frequency as adenomas develop in size and in degree of dysplasia. In particular, two 'hits' are necessary for the development of moderate adenomas [41]. Mutations have been identified in adenomas as small as 3 mm in diameter and this, taken with the observation that the frequency of mutations in adenomas is the same as that in carcinomas, suggests that *APC* mutations are likely to be the initiating step in colorectal carcinogenesis.

The distribution of mutations in sporadic cancers differs from that seen in FAP [40]. Two-thirds of somatic mutations are found in a region of the *APC* gene known as the mutation cluster region (*Figure 6.2*) which represents only 8% of the coding sequence of the gene (codons 1286–1513), a region which has been associated with profuse polyps in FAP patients. However like the germ-line mutations seen, the majority of sporadic mutations are truncating mutations with around 60% being small deletions or insertions causing frameshifts and the majority of the remainder point mutations, primarily C to T transitions [40]. It has been suggested that mutations in *APC* may occur through replication errors in genes such as the mismatch repair genes described below rather than being caused by the action of mutagens in feces since the nature of *APC* mutations is different to those seen in *p53* [42].

6.2.7 Mismatch repair defects

Microsatellite instability, termed replication error positive or RER positive, has been found in around 20% of colorectal cancers (see Chapter 4)[43,44]. RER positive tumors show many of the features of hereditary nonpolyposis colon cancer (HNPCC) lesions in that they are frequently diploid or near diploid and occur in the right side of the colon. Patients whose tumors show microsatellite

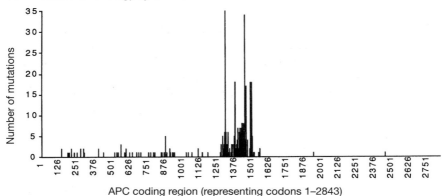

APC coding region (representing codons 1–2843)

Figure 6.2: Distribution of mutations in *APC* in sporadic colorectal cancer. The localization of the majority of mutations in the mutation cluster region (MCR: codons 1286–1513) can be seen. The codon number is shown on the x-axis. Reproduced from ref. 57 with permission from the BMJ Publishing Group.

instability also have increased survival [45]. Preliminary studies suggested that replication errors manifest themselves early in the development of colorectal tumors although others have found no evidence of errors in adenomas. Recently it was shown that the only adenomas with replication errors were those with foci of carcinoma *in situ*, suggesting that the errors occur at the adenoma–carcinoma transition [46]. No difference in the incidence of instability was seen between primary colorectal tumors and liver metastases, suggesting that the changes are not associated with a more aggressive phenotype [47]

As microsatellite instability in tumors from HNPCC patients is caused by mutations in the mismatch repair genes, similar mutations have been sought in cases of sporadic cancers showing instability. Mutations in one of the genes, *hMSH2*, were identified in around 26% of cases [48] and it remains to be determined whether mutations in the other three genes account for the remainder of cases. Mutations in other genes such as polymerase delta may however account for a proportion of cases of instability [49]. In general, patients with instability in their sporadic tumors do not have germline mutations in their mismatch repair genes. One group of patients has however proved to be the exception. These were patients who developed sporadic colon cancer at a relatively young age (<35 years). They showed high levels of instability in 58% of cases and germ-line mutations were identified in almost half of those tested [50]. These findings obviously have implications for genetic testing and subsequent management of their children.

6.2.8   nm23

nm23 was initially isolated because of its association with a low metastatic phenotype in melanoma cell lines and was localized to 17q21, a region which is also commonly deleted in colorectal tumors [51]. There have been conflicting reports on its role in colorectal cancer. In one study, 73% of patients with *nm23* deletions developed metastases after a 25-month follow up compared to only 20% without deletions [52], suggesting that it has a potential role in prognosis. A second study found mutations in *nm23* in 50% of metastatic cancers but in none of those which had not metastasized at the time of surgery [53]. However a third report found no mutations in *nm23* in 26 metastatic, 17 non-metastatic cancers or 43 normal controls [54]. The role of this gene in colorectal cancer currently remains unresolved.

6.2.9   *Other changes*

LOH at chromosome 8p has been a consistent finding in colorectal cancers and has been identified in up to 45% of cases. LOH at this region is seen in only 10% of adenomas suggesting that this involvement is a relatively late event in colorectal carcinogenesis [55].

RB1 has been shown to be amplified in 30% of colorectal cancers and LOH at 13q14 demonstrated in a further 12% [56]. Amplification was particularly associated with aneuploidy. Any prognostic significance in these findings is not yet known.

6.3   Genetic changes in familial colorectal cancer

One method of early detection of colorectal cancer is to identify those at highest risk of developing disease and to target them with the most intensive screening. At the present time this may be a more efficient and cost effective method of screening compared to population screening. Genetic factors are clearly important in the overall increased risk of developing colorectal cancer and evidence of this can be seen from taking family histories. Empiric risks based on family history are given in *Table 6.2*. Genetic susceptibility is likely to be a major contributor to the increased risk to the relatives of affected individuals. However at the present time there are no molecular tests which can be offered to these individuals. Instead, patients at high risk should be offered screening at regular intervals to prevent the development of undiagnosed advanced cancer. However this is different in the 10% of colorectal cancer which is associated with a dominant family history (*Table 6.3*). Here individuals at risk have a 50% chance of developing cancer so careful monitoring is essential. In contrast to the above situation, a causative gene has been isolated for many of these disorders and molecular tests can therefore be used to direct screening appropriately to those at highest risk.

6.3.1   APC

Familial adenomatous polyposis (FAP) is an autosomal dominantly inherited condition predisposing patients to the development of cancer as described in Section 3.3.2. Until recently, screening of at-risk individuals in families with

Table 6.2: Empiric risk of developing colorectal cancer dependent on family history

Affected relative	Lifetime risk
General population	1:50
One first degree relative	1:17
One first degree relative age under 45	1:10
One first and one second degree relative	1:12
Both parents	1:8.5
Two first degree relatives	1:6
Dominant family history	1:2

Table 6.3: Inherited conditions predisposing to colorectal cancer

Familial adenomatous polyposis (FAP)
Hereditary nonpolyposis colon cancer (HNPCC)
    Site-specific colon cancer (Lynch type 1)
    Family cancer syndrome (Lynch type 2)
    Muir Torre syndrome
Turcot syndrome
Juvenile polyposis
Peutz–Jeghers syndrome

FAP depended on regular, usually annual, sigmoidoscopy from the early teens until the patient was in the mid-forties. Patients who develop polyps can then be offered total colectomy or panproctocolectomy with ileorectal anastomosis. Continued monitoring is still required for the rest of their lifetime as polyps develop in the rectal stump and, increasingly, upper gastro-intestinal cancers are being identified. Following the localization of the *APC* gene in 1986 to chromosome 5, identification of 'at-risk' individuals became possible by linkage analysis. This meant that those at high risk could be monitored regularly whereas those at low risk could have their screening either reduced or discon-

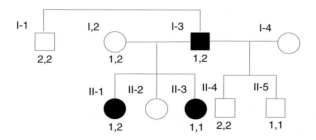

Figure 6.3: Linkage analysis in FAP. This family has been tested with an *Rsa*1 RFLP which lies in exon 11 of the *APC* gene. The disease is associated with allele 1 in this family. Individual II-4 is at low risk of developing FAP so his screening can be reduced or discontinued whereas his brother II-5 is at high risk and regular monitoring by annual sigmoidoscopy will be necessary.

tinued completely depending on the accuracy of the analysis (*Figure 6.3*). However linkage analysis is not possible in all families either because DNA has not been stored from affected individuals before they died or because the disease has arisen as a result of a new mutation, a feature in around 20–25% of FAP cases. Following identification of the *APC* gene, this is no longer a problem as mutation analysis has become possible. However it remains a time consuming process because of the nature and distribution of mutations so that all centers may not be able to offer such a diagnostic service.

Mutations in the *APC* gene vary from one family to the next and around 300 different mutations have so far been identified. They are widely distributed throughout the gene although over 98% of them are located towards the 5' end (*Figure 6.4*). There is therefore a difference from the distribution seen in sporadic colon cancer described above. In common with sporadic cancer, approximately half of the mutations are point mutations with the remainder either small insertions or deletions [57]. The vast majority of these mutations are truncating which means that a technique such as the protein truncation test (PTT) can be used to identify them more easily (see Chapter 13). The ability to detect the causative mutation in families means that accurate presymptomatic diagnosis is now possible (*Figure 6.5*)

Correlations have been made between the position of the mutation and the phenotype (see Section 3.3.2). The presence of congenital hypertrophy of the retinal pigment epithelium (CHRPE) is associated with mutations between exon 9 and codon 1444 in exon 15 [57]. The attenuated form of FAP, which is associated with a later onset of disease and fewer polyps, is associated with mutations 5' to exon 4 [57]. In contrast a severe phenotype has been associated with mutations at codon 1309 and desmoid tumors in particular are more

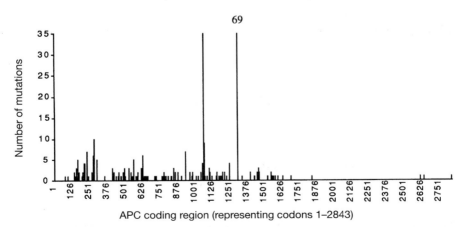

Figure 6.4: Distribution of mutations in *APC* in FAP patients. The two main 'hot spots' for mutations at codons 1061 and 1309 can be seen. The codon number is shown on the x-axis. Reproduced from ref. 57 with permission from the BMJ Publishing Group.

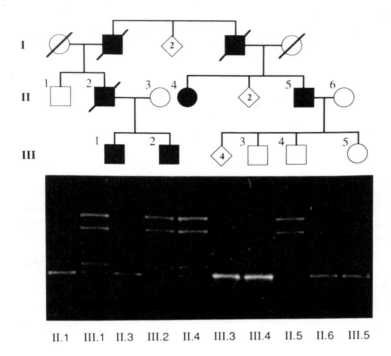

Figure 6.5: Mutation analysis of *APC* by denaturing gradient gel electrophoresis (DGGE) in a family with FAP. The upper panel shows the family pedigree. In the lower panel the results of DGGE analysis are shown. Affected individuals III-1, III-2, II-4 and II-5 show a variant pattern characteristic of the presence of a mutation (subsequently sequenced and shown to be a C to T transition at codon 283). Individuals at risk in this pedigree (III-3, III-4 and III -5) show only a single band, indicating that they have not inherited the mutation present in their father. Reproduced from ref. 57 with permission from the BMJ Publishing Group.

likely to occur in patients with mutations after codon 1444. The explanation for the more severe phenotype is suggested to be due to the fact that mutant APC protein is capable of interacting with the wild-type protein and may therefore have a dominant negative effect. In contrast, short truncated proteins, such as those predicted to occur with mutations seen in attenuated FAP, are unstable and are rapidly degraded. They are therefore incapable of interacting with wild-type protein. The milder phenotype is hence due to a decrease in the level, but not the complete removal, of APC protein.

6.3.2   HNPCC

HNPCC, like FAP, is an autosomal dominantly inherited condition so offspring with an affected parent are again at 50% risk of developing the disease. Tumors develop later in life than in FAP, with a mean age of diagnosis at around 40–45 years. As mentioned in Section 6.2.7, tumors are predominantly found in the proximal rather than the distal colon and synchronous and metachronous tumors are seen in 20–30% of patients. The disease is subdivided into three conditions:

(1) site-specific colorectal cancer or Lynch type 1, in which, as the name suggests, the tumors seen are all found in the colon;
(2) family cancer syndrome or Lynch type 2 which includes endometrial, ovarian, hepatobiliary, small bowel and transitional cell cancer as well as colon cancer;
(3) Muir Torre syndrome in which sebaceous tumors are found in addition to those found in Lynch type 2.

Unlike FAP, the diagnosis cannot be made on an individual patient but is based instead on family history. The criteria to make a diagnosis of HNPCC have been termed 'the Amsterdam criteria' and include:

(1) at least three relatives must have histologically proven cases of colon cancer;
(2) one relative must be the first degree relative of another;
(3) at least two generations should be affected;
(4) cancer should have been diagnosed in one individual before the age of 50;
(5) FAP should be excluded.

Until 1993, monitoring of individuals at risk depended on regular surveillance by colonoscopy or ultrasound as appropriate. Following the cloning of the genes responsible for this condition (*hMSH2*, *hMLH1*, *hPMS1* and *hPMS2*), linkage analysis and mutation analysis have become possible to identify high risk individuals.

Linkage analysis in HNPCC is more complicated than in FAP due to genetic heterogeneity. Larger pedigrees are needed so that linkage to one of the four causative genes can convincingly be proven since it cannot be assumed which locus is responsible for the disease. This type of analysis is more limited but has been carried out in many families (*Figure 6.6*). Linkage analysis has shown that *hMSH2* and *hMLH1* account for the majority of cases of HNPCC at approximately equal frequency with the other two genes accounting for less than 10% [58]. As with FAP, now that the genes have been identified, direct mutation analysis is possible but it is an even more difficult task than for *APC* as four genes have to be investigated rather than one.

As with *APC*, mutations are widely distributed throughout the causative genes with the majority of cases being point mutations or small deletions and insertions. Although many mutations are truncating, meaning that the PTT can be used as an initial screening test, there is a much larger proportion of missense mutations in *hMSH2* and *hMLH1* than in *APC*. There are to date no obvi-

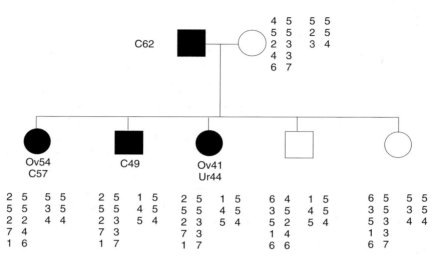

Figure 6.6: Linkage analysis in HNPCC. The family has been tested with five microsatellites linked to *hMSH2* (left hand group under each individual) and with three markers linked to *hMLH1* (right hand group). All three affected members of the family share the same haplotype with the markers linked to *hMSH2* (2,5,2,7,1) which must have been inherited from their affected father. They have inherited different chromosome 3 haplotypes so the possibility of the disease being caused by an *hMLH1* mutation can be excluded. The two unaffected sibs have inherited the opposite paternal haplotype with the chromosome 2 markers so are at low risk of developing the disease. Pedigree courtesy of Dr N. Froggat, Molecular Genetics, Addenbrookes Hospital, Cambridge.

ous hot spots for mutations. Although there have been two reports of mutations found in several families these cannot be considered to be common mutations. In one case two common mutations, accounting for around 63% of HNPCC families in Finland, could be traced back to two common ancestors [59]. In the other, an exon 5 splice site mutation was found in four out of 33 British families [60]. Unfortunately, analysis of further families suggests that the mutation is not as common as it initially appeared. A recent report has suggested that mutations in *hMLH1* are primarily clustered in the region encompassing exons 15 and 16 but larger studies are required to confirm this [61].

There appears to be no correlation between the nature of the mutation and the spectrum of tumors seen in the families. For example, two mutations in exon 12 of *hMSH2* in families with Muir Torre syndrome have been described (*Figure 6.7*) but in other families exon 12 mutations have resulted in a classical HNPCC phenotype [62]. This suggests that other modifier genes may play a part in the clinical phenotype as well as the possible interaction of environmental factors.

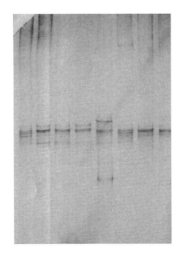

Figure 6.7: Mutation analysis in a family with Muir Torre syndrome. The figure shows SSCP analysis of exon 12 of *hMSH2*. Family members in lanes 1 and 6 show the pattern characteristic of the mutation (subsequently sequenced and shown to be a 2 bp deletion causing a frameshift). At-risk individuals (lanes 2, 3, 4, 7 and 8) show the normal pattern and are therefore not likely to develop Muir Torre syndrome. The individual in lane 5 comes from another HNPCC family and this pattern was subsequently shown to be caused by a 5 bp deletion in this exon.

6.3.3 Turcot syndrome

This hereditary condition was originally described as a syndrome of colorectal polyposis plus a primary tumor of the central nervous system, the mode of inheritance of which has been controversial. A recent study has shown that this condition can result from two different types of germ-line mutations; either mutations in *APC* or in one of the mismatch repair genes. It is therefore clear that brain tumors can be included in the range of tumors seen in both FAP and HNPCC. This has additional implications for the care of these patients [63].

References

1. Parker, S.L., Tong, T., Bolden, S.B. and Wingo, P.A. (1996) *CA – Cancer J. Clinic.*, **46**, 5.
2. Morson B.(1974) *Cancer*, **34**, 845.
3. Nomura, A. (1990) *J. Natl Cancer Inst.*, **82**, 894.
4. Field, J.K. and Spandidos, D.A. (1990) *Anticancer Res.*, **10**, 1.
5. Smith, D.R., Myint, T. and Goh, H.-S. (1993) *Br. J. Cancer*, **68**, 407.
6. Imaseki, H., Hayashi, H., Taira, M., Ito, Y., Tabata, Y., Onada, S., Isono, K. and Tatibana, M. (1989) *Cancer*, **64**, 704.
7. Pavelic, Z.P., Pavelic, L., Kuvelkar, R. and Gapany, S.R. (1992) *Anticancer Res.*, **12**, 171.
8. Astrin, S.M. and Costanzi, C. (1989) *Sem. Oncol.*, **16**, 138.
9. Rodrigues-Alfageme, C., Stanbridge, E.J. and Astrin, S.M. (1992) *Proc. Natl Acad. Sci. USA*, **89**, 1482.
10. Bos, J.L., Fearon, E.R., Hamilton, S.R., Verlaan-de Fries, M., van Boom, J.H., van der Eb, A.J. and Vogelstein, B. (1987) *Nature*, **327**, 293.
11. Fearon, E.R. and Vogelstein, B. (1990) *Cell*, **61**, 759.
12. Minamoto, T., Ronai, Z., Yamashita, N., *et al.* (1994) *Int. J. Oncol.*, **4**, 397.
13. N. Yamashita, T. Minamoto, A. Ochiai, M. Onda and H. Esumi (1995) *Cancer*,

75, 1527.

14. Laurent Puig, P., Olschwang, S., Lelattre, O. *et al.* (1992) *Gastroenterology*, **102**, 1136.
15. Bell, S., Scott, N., Cross, D. *et al.* (1993) *Gastroenterology*, **104**, 57.
16. Finkelstein, S.D., Sayegh, R., Bakker, A., Swalensky, P. (1993) *Arch. Surg.*, **28**, 526.
17. Moerkerk, P., Arends, J.W., van Driel, M., de Bruine, A., Goeij, A., ten Kate, J (1994) *Cancer Res.*, **54**, 3376.
18. Sidransky, D., Tukino, T., Hamilton, S.R. Vogelstein, B. (1992) *Science*, **256**, 102.
19. Smith-Ravine, J., England, J., Talbot, I.C. and Bodmer, W. (1995) *Gut*, **36**, 81.
20. Baker, S.J., Preisinger, A.C., Jessup, M. *et al.* (1990) *Cancer Res.*, **50**, 7717.
21. Kern, S.E., Fearon, E.R., Tersmette, K.W.F., Enterline, J.P., Leppart, M. and Nakamura, Y. (1989) *J. Am. Med. Assoc.* **261**, 3099.
22. Khine, K., Smith, D. Hak-Su, G. *et al.* (1994) *Cancer*, **73**, 28.
23. Remvikos, Y., Tominaga, O., Hammel, P., Laurent-Puig, P., Salmon, R.J., Dutrillaux, B. and Thomas, G. (1992) *Br. J. Cancer*, **66**, 758.
24. Nathanson, S.D., Linden, M.D., Tender, P., Zarbo, R.J., Jacobson, G. and Nelson, L.S. (1994) *Dis. Colon Rectum*, **37**, 527.
25. Hall, P.A. and Lane, D. (1994) *J. Pathol.*, **172**, 1.
26. Bass, I.O., Mulder, J.W.R., Offerhaus, G.F.A. Vogelstein, B. and Hamilton, S. (1994) *J. Pathol.*, **172**, 5.
27. Mulder, J.W.R., Bass, I.O., Polak, M.M., Goodman, S.N. and Offerhaus, G.J.A. (1995) *Br. J. Cancer*, **71**, 1257.
28. Dix, B., Robbins, P., Carrello, S., House, A. and Iacopetta, B. (1994) *Br. J. Cancer*, **70**, 585.
29. Sun X.-F., Carstensen, J.M., Zhang, H., Stal, O., Wingern, S., Hatschek, T. and Nordenskjold, B. (1992) *Lancet*, **340**, 1369.
30. Hamelin, R., Laurent-Puig, P., Olschwang, S. *et al.* (1994) *Gastroenterology*, **108**, 42.
31. Hasegawa, H., Ueda, M., Furukawa, K., Watanabe, M., Teramoto, T., Mukai, M. and Kitajima, M. (1995) *Int. J. Cancer*, **64**, 47.
32. Carder, P.J., Cripps, K.J., Morris, R., Collins, S., White, S., Bird, C.C. and Wyllie, A.H. (1995) *Br. J. Cancer*, **71**, 215.
33. Kinzler, K., Nilbert, M.C., Vogelstein, B., Bryan, T.M., Levy, D.B., Smith, K.J. (1991) *Science*, **251**, 1366.
34. Curtis, L.J., Bubb, V.J., Gledhill, S., Morris, R.G., Bird, C.C. and Wyllie, A.H. (1994) *Hum. Mol. Genet.*, **3**, 443.
35. Fearon, E.R., Cho, K.R., Nigro, J.M. *et al.* (1990) *Science*, **247**, 49.
36. Cho, K.R. and Fearon, E.R. (1995) *Curr. Opin. Genet. Dev.*, **5**, 72.
37. Jen, J., Kim, H., Piantadosi, S. *et al.* (1994) *N. Engl. J. Med.*, **331**, 213.
38. Zetter, B.R. (1993) *Sem. Cancer Biol.*, **4**, 219.
39. Vogelstein, B., Fearon, E.R., Hamilton, S.R. *et al.* (1988) *N. Engl. J. Med.*, **319**, 525.
40. Nagase, H. and Nakamura, Y. (1993) *Hum. Mut.*, **2**, 425.
41. Ichii, S., Hori, A., Nakatsuru, S., Furuyama, J., Utsonomiya J. and Nakamura, Y. (1993) *Hum. Mol. Genet.*, **1**, 387.
42. Miyaki, M. Konishi, M., Kikuchi-Yanoshita, R., Enomoto, M., Igari, T., Tanaka, K. (1994) *Cancer Res.*, **54**, 3011.
43. Aaltonen, I.A., Peltomaki, P., Leach, F.S., Sistonen, P., Pylkkanen, L. and Mecklin, J.-P. (1993) *Science*, **260**, 812.

44. Thibodeau, S.N., Bren, G. and Schaid, D. (1993) *Science*, **260**, 816.
45. Lothe, R.A., Peltomaki, P., Meling, G.I., Aaltonen, L.A., Nystrom-Lahti, M. and Pylkkanen, L. (1993) *Cancer Res*, **53**, 5849.
46. Young, J., Searle, J., Buttenshaw, R. *et al.* (1995) *Genes Chrom. Cancer*, **12**, 251.
47. Ishimaru, G., Adachi, J., Shiseki, M., Yamaguchi, N., Muto, T. and Yokota, J. (1995) *Int. J. Cancer*, **64**, 153.
48. Borrensen, A-L., Lothe, R.A., Meling, G.I. *et al.* (1995) *Hum. Mol. Genet.*, **4**, 2065.
49. Costa, L.T., Liu, B., El-Deiry, W.S. *et al.* (1995) *Nature Genetics*, **9**, 9.
50. Liu, B., Farrington, S.M., Petersen, G.M. *et al.* (1995) *Nature Med.*, **1**, 348.
51. Purdie, C.A., Piris, J., Bird, C.C. and Wyllie, A.H. (1995) *J. Pathol*, **175**, 297.
52. Cohn, K.H., Wang F., Desoto-Paix, F. *et al.* (1991) *Lance*t, **338**, 722.
53. Wang, L., Patel, U., Ghosh, L., Chen, H., and Banerjee, S. (1993) *Cancer Res.*, **53**, 717.
54. Myeroff, L. and Markowitz, S.D. (1993) *J. Natl Cancer Inst.*, **85**, 147.
55. Cunningham, C., Dunlop, M.G., Bird, C. C. and Wyllie, A.H. (1994) *Br. J. Cancer*, **70**, 18.
56. Meling, G.I., Lothe, R.A., Borresen, A.-L., Hauge, S., Graue, C., Clausen , O.P.F. and Rognum, T.O.(1991) *Br. J. Cancer*, **64**, 475.
57. Wallis, Y. and Macdonald, F. (1996) *J. Clin. Mol. Pathol.*, **49**, M65.
58. Nystrom-Lahti, M., Parsons, R., Sistonen, P., *et al.* (1994) *Am. J. Hum. Genet.*, **55**, 659.
59. Nystrom-Lahti, M., Kristo, P., Nicolaides, N.C., Chang, S-Y., Aaltonen, L.A., Moisio, A-L. *et al.* (1995) *Nature Med.*, **1**, 1203.
60. Froggat, N.J., Joyce, J.A., Davies, R., Evans, D.G.R., Ponder, B.A.J., Barton, D. and Maher, E. (1995) *Lancet*, **345**, 727.
61. Wijnen, Khan, P.M., Vasen, H. *et al.* (1996) *Am. J. Hum. Genet.*, **58**, 300–307.
62. Kolodner, R.D., Hall, N.R., Lipford,J. *et al.* (1995) *Genomics*, **24**, 516.
63. Hamilton, S.R., Liu, B., Parsons, R.E. *et al.* (1995) *N. Engl J. Med.*, **332**, 839.

Chapter 7

Gastrointestinal cancers

7.1 Gastric cancers

The incidence of gastric cancer has decreased in the Western world but is a major cause of death from cancer in other areas such as Japan and South America. The age-adjusted death rates in the major industrialized countries of the world are shown in *Table 7.1* [1] Approximately 12 000 new cases are seen in the UK annually. With the advent of screening programs, particularly endoscopic examination, there has been an increase in the detection of early disease. However the 5-year survival rate remains low at around 11%.

Table 7.1: Age-adjusted death rates from gastric cancer in the major industrialized countries of the world

Country	Age-adjusted death rates (1990–1993) per 100 000 population	
	Male	Female
Japan	32.8	14.2
Italy[a]	16.9	7.9
Germany	13.9	7.3
UK[c]	11.8	4.8
France[b]	8.2	3.3
Canada[b]	7.1	3.2
USA[b]	5.0	2.3

Abstracted from ref. 1.
[a]1990–1991; [b]1990–1992; [c]1992–1993.

Many different histological classifications have been devised for gastric cancer but the most commonly used is that of Lauren who divided the disease into two groups, intestinal or diffuse [2]. These two types differ not only morphologically but also show differences in their clinical progress and in their epidemiology. The intestinal form of gastric cancer is thought to develop via a stepwise process from chronic to atrophic gastritis, intestinal metaplasia (particularly type 3), dysplasia and finally carcinoma. However there is no obvious sequential pattern of changes for the diffuse type. Some families have been described in whom gastric cancer appears to segregate as an autosomal dominant disease and gastric cancer is associated with FAP and Lynch type 2

99

Cancer Res., **50**, 4911.
8.   Lemoine, N.R., Jain, S, and Silvestre, F. (1991) *Br. J. Cancer*, **64**, 79.
9.   Tahara, E., Sumiyoshi, H., Hata, J., Yasui, W., Taniyama, K., Hayashi, T., Nagae, S. and Sakamoto, S. (1986) *Jpn. J. Cancer Res.*, **77**, 145.
10. Kameda, T., Yasui, W., Yoshida, K., *et al.* (1990) *Cancer Res.*, **50**, 8002.
11. Jain, S., Filipe, M.I., Gullick, W.J., Linehan, J. and Morris, R.W. (1991) *Int. J. Cancer*, **48**, 668.
12. Nakatsuru, S., Yanagisawa, A., Ichii, S., Tahara, E., Kato, Y., Nakamura, Y. and Horii, A. (1993) *Hum. Mol. Genet.*, **1**, 559.
13. Nakatsuru, S., Yanagisawa, A., Furukawa, Y., Ichii, S., Kato, Y., Nakamura, Y. and Horii, A. (1993) *Hum. Mol. Genet.*, **2**, 1463.
14. Tamura, G., Maesawa, C., Suzuki, Y., *et al.* (1994) *Cancer Res.*, **54**, 1149.
15. Yamada, Y., Yoshida, T., Hayashi, K., *et al.* (1991) *Cancer Res.*, **51**, 5800.
16. Bartek, J., Bartkova J., Vojtesek, B. *et al.* (1991) *Oncogene*, **6**, 1699.
17. Martin, H.M., Filipe, M.I., Morris, R.W., Lane, D.W and Silvestre, F. (1992) *Int. J. Cancer*, **50**, 859.
18. Kakeji, Y., Korenaga, D., Tsujitani, S., Baba, H., Anai, H., Maehara, Y. and Sugimachi, K. (1993) *Br. J. Cancer*, **67**, 589.
19. Joypaul, B.V., Hopwood, D., Newman, E.C., Qureshi, S., Grant, A., Ogston, S.A., Lane, D.P and Cuschieri, A. (1994) *Br. J. Cancer*, **69**, 943.
20. Craanen, M.E., Blok, P., Dekker, W., Offerhaus, G.J.A. and Tytgat, G.N.J. (1995) *Gut*, **36**, 848.
21. Renault, B., Brock, M. v.d., Fodde, R., Wijnen, J., Pellagata, N.S., Amadori, D., Khan, P.M. and Ranzani, G.N. (1993) *Cancer Res.*, **53**, 2614.
22. Han, H.-J., Yanagisawa, A., Kato, Y., Park, J.-G. and Nakamura, Y. (1993) *Cancer Res.*, **53**, 5087.
23. Dos Santos, N.R., Seruca, R., Constancia, M., Seixas, M. and Sobrinho-Simoes, M. (1996) *Gastroenterology*, **110**, 38.
24. Gaudray, P., Szepetowski, P., Escot, C., Birnbaum, D. and Theillet, C. (1992) *Mutat. Res.*, **276**, 317.
25. Kitagawa, Y., Ueda, M., Ando, N., Shinozawa, Y., Shimizu, N. and Abe, D. (1991) *Cancer Res.*, **51**, 1504.
26. Jiang, W., Zhang, Y.J., Kahn, S.M. *et al.* (1993) *Proc. Natl Acad. Sci. USA*, **90**, 9026.
27. Adelaide, J., Monges, G., Derderian, C., Seitz, J.-F. and Birnbaum, D. (1995) *Br. J. Cancer*, **71**, 64.
28. Sarbia, M., Porschen, R., Borchard, F. *et al.* (1994) *Cancer*, **74**, 2218.
29. Wang, D.Y., Xiang, Y.Y., Tanaka, M., *et al.* (1994) *Cancer*, **74**, 3089.
30. Casson, A.G., Kerkvliet, N. and O'Malley, F. (1995) *J. Surg. Oncol.*, **60**, 5.
31. Vijeyasingam, R., Darnton, S.J., Jenner, K., Allen, C.A., Billingham, C. and Mathews, H. (1994) *Br. J. Surg.*, **81**,1623.
32. Ramel, S., Reid, B.J., Sanchez, C.A. *et al.* (1992) *Gastroenterology* **102**, 1589.
33. Younes, M., Lebovitz, R.M., Lechago, L.V. and Lechago, J. (1993) *Gastroenterology*, **105**, 1637.
34. Rice, T.W., Goldblum, J.R., Falk, G.W., Tubbs, R.R., Kirby, T.J. and Casey, G. (1994) *J Thorac. Cardiovasc. Surg.*, **108**, 1132.
35. Hollstein, M.C., Metcalf, R.A., Welsh, J.A., Montesano, R., Harris, C.C. (1990) *Proc. Natl Acad. Sci. USA*, **87**, 9958.
36. Casson, A.G., Mukhopadhyay, T., Cleary, K.R. *et al.* (1991) *Cancer Res.*, **51**, 4495.

37. Neshat, K., Sanchez, C.A., Galipeau, P.C., Blount, P.L., Levine, D.S., Josyln, G. and Reid, B. (1994) *Gastroenterology*, **106**, 1589.
38. Schneider, P.M., Casson, A.G., Levin, B. *et al.* (1996) *J Thorac. Cardiovasc. Surg.*, **111**, 323.
39. Casson, A.G., Manolopoulos, B., Troster, M., Kerkvliet, O'Malley, F., Inculet, R. and Finley, R. (1994) *Am. J. Surg.*, **167**, 52.
40. Huang, Y., Melzer, S.J., Yin, J. *et al.* (1993) *Cancer Res.*, **53**, 1889.
41. Ozawa, S., Ueda, S., Ando, N. *et al.* (1987) *Int. J. Cancer*, **39**, 333.
42. Lu, S-H., Hsieh, L.L., Luo, F.C. and Weinstein, I.B. (1988) *Int. J. Cancer*, **42**, 502.
43. Al-Kasspooles, R.K., Moore, J.H., Orringer, M.B. and Beer, D.G. (1993) *Int. J. Cancer*, **54**, 213.
44. Iahara, K., Shiozaki, H., Tahara, H. *et al.* (1993) *Cancer*, **71**, 2902.
45. Jankowski, J., Coghill, G., Hopwood, D. and Wormsley, K.G. (1992) *Gut*, **33**, 1033.
46. Mori, T., Miura, K., Aoki, T., Nishihira, T., Mori, S. and Nakamura, Y. (1994) *Cancer Res.*, **54**, 5269.
47. Igaki, H., Sasaki, H., Tachimori, Y. *et al.* (1995) *Cancer Res.*, **55**, 3421.
48. Melzer, S.J., Yin, J., Manin, B. *et al.* (1994) *Cancer Res.*, **54**, 3379.
49. Hruban, R.H., van Mansfield, A.D.M., Offerhaus, G.J.A. *et al.* (1993) *Am. J. Pathol.*, **143**, 545.
50. Tada, M., Omata, M., Kawai, S., Saisho, H.,Ohto, M., Saiki, R.K. and Sninsky, J.J. (1993) *Cancer Res.*, **53**, 2472.
51. Watanabe, H., Sawabu, N., Ohta, H. *et al.* (1993) *Jpn. J. Cancer Res.*, **84**, 961.
52. Van Laethern, J-L., Vertongen, P., Deviere, J., Van Ramelbergh, J., Rickaert, F., Cremer, M. and Robberecht, P. (1995) *Gut*, **36**, 781.
53. Iguchi, H., Sugano, K., Fukayama, N. *et al.* (1996) *Gastroenterology*, **110**, 221.
54. Caldas, C., Hanh, S.A., Hruban, P.H., Redston, M.S., Yeo, C.J. and Kern, S.E. (1994) *Cancer Res.*, **54**, 3568.
55. Berthelemy, P., Bouisson, M., Escorrou, J., Vaysse, N., Rumeau, J.L and Pradayrol, L. (1995) *Ann. Intern. Med.*, **123**, 188.
56. Yanagisawa, M., Ohtake, K., Ohashi, K., Hori, M., Kitagawa, T., Sugano, H. and Kato, Y. (1993) *Cancer Res.*, **53**, 953.
57. Tada, M., Ohashi, M., Shiratori, Y. *et al.* (1996) *Gastroenterology*, **110**, 227.
58. Lemoine, N.R., Hughes, C.M., Barton, C.M. *et al.* (1992) *J. Pathol.*, **166**, 7.
59. Barton, C.M., Hall, P.A., Hughes, C.M. *et al.* (1991) *J. Pathol.*, **163**, 111.
60. Hall, P.A., Hughes, C.M., Staddon, S.L. *et al.* (1990) *J. Pathol.*, **161**, 195.
61. Lemoine, N.R., Lobresco, M., Leung, H. *et al.* (1992) *J. Pathol.*, **168**, 269.
62. Barton, C.M., Staddon, S.L., Hughes, C.M. *et al.* (1991) *Br. J. Cancer*, **64**, 1076.
63. Simon, B., Weinel, R., Hohne, M., Watz, J., Schmidt, J., Kortner, G. and Arnold, R. (1994) *Gastroenterology*, **106**, 1645.
64. Kalthoff, H., Schmiegel, W., Roeder, C. *et al.* (1993) *Oncogene*, **8**, 289.
65. Scarpa, A., Capelli, P., Mukai, K. *et al.* (1993) *Am. J. Pathol*, **142**, 1534.
66. Hohne, M.W., Halatsch, M.E., Kahl, G.F. and Weinel R.J. (1992) *Cancer Res.*, **52**, 2616–2619.
67. Caldas, C., Hahn, S.A., da Costa, L.T., Redston, M.S., Schutte, M. and Seymour, A.B. (1994) *Nature Genetics*, **8**, 27.
68. Wang, J., Zindy, F., Chenivesse, X., Lamas, E., Henglein, B. and Brechot, C. (1992) *Oncogene*, **7**, 1653.
69. Ogata,N., Kamimura, T. and Asakura, H. (1991) *Hepatology*, **13**, 31.

70. Zhang, X.-K., Huang, D.-P., Qiu, D.-K. and Chiu, J.-F. (1990) *Oncogene*, **5**, 909.
71. Gu, J.-R., Hu, L.-F., Cheng, Y.-C. and Wan, D.-F. (1986) *J. Cell. Physiol.* **4** (Suppl.), 13.
72. Bressac, B., Kew, M., Wands, J. and Ozturk, M. (1991) *Nature*, **350**, 429.
73. Hsu, I.C., Metcalf, R.A., Sun, T., Welsh, J.A., Wang, N.J. and Harris, C.C. (1991) *Nature*, **350**, 427.
74. Ozturk, M., Bressac, B., Puisieux, A. *et al.* (1991) *Lancet*, **338**, 1356.
75. Harris, C.C. (1996) *Br. J. Cancer*, **73**, 261.
76. Wang, X.E., Yeh, H., Schaeffer, L., Roy, R., Moncollin, V. and Egly, J.M. (1995) *Nature Genetics*, **10**, 188.
77. Tsuda, H., Zhang, W., Shimosato, Y. *et al.* (1990) *Proc. Natl Acad. Sci. USA*, **87**, 6791.
78. Voravud, N., Foster, C.S., Gilbertson, J.A., Sikora, K. and Waxman, J. (1989) *Hum. Pathol.*, **20**, 1163.
79. Ohashi, K., Nakajima, Y., Kanehiro, H., Tsutsumi, M., Taki, J., Aomatsu, Y. (1995) *Gastroenterology*, **109**, 1612.
80. Hayward, N.K., Little, M.H., Mortimer, R.H., Clouston, W.M. and Smith, P.J. (1988) *Cancer Cell Cytogenet.*, **23**, 95.

Chapter 8

Breast cancer

8.1 Introduction

Breast cancer is one of the most common cancers in women in the developed
countries of the world and it is the cause of death in approximately 20% of all
females who die from cancer in these countries. Survival figures have not
altered significantly over the years. The age-adjusted death rates for women in
the major industrialized countries are shown in *Table 8.1*.

Table 8.1: Age-adjusted death rates in the major industrialized countries of the world

Country	Age-adjusted death rates (1990–1993) per 100 000 population
UK[c]	27.7
Canada[b]	23.0
Germany	22.2
Italy[a]	20.7
USA[b]	22.0
France[b]	19.7
Japan	6.6

Abstracted from ref. 1.
[a]1990–1991; [b]1990–1992; [c]1992–1993.

There are marked differences in the incidence of breast cancer in different
places, the predominant impression being that the disease is more common
among Caucasians living in the colder climates and more highly industrialized
countries of the Western hemisphere [2]. However, a recent survey has shown
that, with the possible exception of China, breast cancer incidence rates have
been increasing over the last 20 years in all age groups in all countries of the
world for which rates are obtainable [3].

The tumors are classified as noninvasive or invasive, the majority (76%)
belonging to the group of invasive ductal breast cancers. Patients can be staged
using the clinical characteristics of tumor diameter (T), lymph node involvement (N) and the presence of distant metastases (M), hence the nomenclature

117

TNM staging. Prognostic factors include lymph node status, degree of differentiation of the tumor and the presence of estrogen and/or progesterone receptors. The DNA content (ploidy) of tumors has been suggested as a prognostic indicator but its use is limited at present. However, DNA analysis of *in situ* ductal carcinomas of the breast by flow cytometry has indicated that major genetic alterations and DNA heterogeneity are early events in carcinogenesis and that they are already established at the preinvasive stage [4].

A number of risk factors have been associated with development of the disease, including cigarette smoking, alcohol consumption, age at menarche, age at first childbirth, fat in the diet and a family history [5].

8.2   Genetic changes in breast cancer

Although relatively little is known about the molecular mechanisms leading to breast cancer development, breast cancers have probably been studied more than any other tumor type with regard to oncogene expression. *MYC*, *ERBB2* or one of the *RAS* family have been found to be expressed in over 60% of cases.

8.2.1   MYC

Very few examples of rearrangements of the *NMYC* or *LMYC* genes have been found in breast cancer. On the other hand, there is considerable evidence for *MYC* amplification although the reported incidence varies from 4 to 41% [6]. Three large studies of *MYC* amplification in breast cancer have produced different conclusions as to the value of this marker in prognosis [7].

In one study, including 80 primary breast carcinomas as well as benign tissues and nodal metastases, amplification was the major alteration in the *MYC* gene occurring in 18% of tumors, and one example of a rearrangement was also seen. Amplification was primarily associated with infiltrating ductal carcinomas and poorly differentiated tumors. No stage I tumors showed any abnormalities in *MYC*. A significant correlation was seen between amplification of *MYC* and poor short-term prognosis, implicating this gene in the progression of breast cancer. In a second study of tumors from 121 patients, over 30% had abnormalities in *MYC*. There was a significant correlation between tumors of post-menopausal patients and amplification of *MYC*, but no correlation with tumor grade, receptor status or presence of metastatic disease was found. No survival rates were presented in this study. Finally, a prospective study of 125 patients found amplification of *MYC* in 18% of cases but no clinical correlations were detected, except a highly significant association between *MYC* amplification and inflammatory carcinomas, suggesting that *MYC* might contribute to the rapid progression of this subtype of breast cancer. Increased levels of *MYC* mRNA have also been found in breast cancer and have not always correlated with gene amplification. In one study almost 50% of tumors had elevated mRNA levels and this correlated with the presence of lymph node metastases. Amplification was seen in only 11% of tumors [7].

Monoclonal antibodies Myc1-6E10 and Myc1-9E10 have both been used to study p62 levels in breast cancers but have not clarified the role of *MYC* in this disease. Seventy percent of cases in one study, using antibody Myc1-9E10, showed expression of p62 in both the nucleus and occasionally in the cytoplasm, but there was no correlation with histopathological grade, presence of metastases in the lymph nodes, or receptor status [7]. This antibody may not be of value for detecting abnormalities in the *MYC* gene, because discrepancies have been demonstrated between the levels of MYC protein, as determined immunohistochemically, and the levels of mRNA measured by *in situ* hybridization. Neither of these techniques could be used to indicate the presence of amplification [8]. These results suggest that there may be problems in using measurement of oncoprotein levels with monoclonal antibodies in routinely processed tissues as a measure of abnormalities in the oncogene itself.

Finally, a sensitive enzyme assay has been used as a means of measuring *MYC* oncoprotein expression in tissue extracts. All tumors had elevated levels of p62 compared with normal tissue, but no correlation with age, nodal status, receptor status, histopathological grade or survival was observed. There was, however, an association between tumor size (T) and high levels of MYC oncoprotein expression [7].

The data on the prognostic value of *MYC* amplification are therefore conflicting, with some studies suggesting it is an independent powerful prognostic factor, particularly in node-negative and steroid receptor-positive cancers [9], and others suggesting that it is not a suitable prognostic marker for routine purposes [10]. Taken together these results do not show any clearly defined diagnostic or prognostic role for *MYC* in breast cancer at the present time, and any role it does have is likely to be complex.

8.2.2 ERBB2 *and* RAS

More useful information has been obtained from measurements of *ERBB2* and the *RAS* gene family. The majority of studies have detected elevated levels of *HRAS* mRNA in primary breast carcinomas. Elevated expression was associated with advanced histological types. In contrast, a further study detected elevated expression of *NRAS* and *KRAS*, but only one tumor showed any increase in *HRAS* [7].

Increased expression of RAS p21 has been detected in 63–83% of malignant breast tumors compared with low levels in benign tissues. A single study was unable to detect differences in RAS p21 expression between malignant and benign breast tumors. In general there are no significant clinical correlations, although one study demonstrated higher RAS p21 expression in carcinomas from post-menopausal women. A direct-binding quantitative competition radioimmunoassay has been developed for RAS p21 using monoclonal antibody Y13 259 to determine absolute levels of the protein. Again, approximately two-thirds of carcinomas demonstrated high levels of the protein compared with lactating breast and benign tissues. Recent studies using this technique demonstrated higher levels of p21 in post-menopausal patients. Levels of RAS p21 have also been determined in breast tissue using Western blotting techniques and the results have confirmed the immunohistochemical findings of high levels of the protein in malignant tissues. An association was also found between high levels of *RAS* p21 and extent of tumor and a significant correlation was seen between high levels and short disease-free interval.

Studies of the *RAS* family in breast cancer have therefore found little evidence of point mutations or structural alterations in the genes, but high levels of *RAS* appear to be associated with progression of breast cancer and poor prognosis [7].

As in lung cancer, polymorphisms associated with *HRAS* have been investigated in breast cancer. In one study, four common alleles and 16 rare alleles were found in the normal and cancer-bearing populations. The frequency of two particular common alleles was diminished in the breast cancer patients with concomitant increases in two rare alleles. One of these rare alleles was significantly associated with breast cancer and is potentially of use in risk analysis. In a second study, *HRAS* allele loss has been found in breast tumors and correlates significantly with grade III tumors, lack of estrogen/progesterone receptors and the presence of distant metastases. A third study has confirmed that allele loss correlates with low levels of estrogen receptor [7]. The association of *HRAS* polymorphisms and breast cancer has now been submitted to meta-analysis and this has shown that women with a single copy of the rare allele have a 1.7-fold increased risk of breast cancer compared to the general population, while those who are homozygous for the rare allele have a 4.6-fold increased risk [11].

ERBB2 amplification in breast cancer has been demonstrated by many investigators, in frequencies ranging from 10 to 30%. In a preliminary study, amplification of *ERBB2* correlated with poor prognosis in lymph node positive patients. Several subsequent studies confirmed this finding, but others have been unable to show any correlation. The workers who produced the initial results have corroborated their findings in a large study of 526 patients [12] and two other groups have found similar results in lymph node negative breast cancer. In one study co-amplification of *ERBB2* and *ERBA1* was found in 23% of breast tumors and was shown to be a strong indicator of metastatic potential.

Antibodies to the ERBB2 protein have also been used to study overexpression of *ERBB2* in breast cancer and there are several reports which correlate the levels of ERBB2 protein with gene amplification. As with direct analysis of gene amplification, immunostaining for ERBB2 protein as a prognostic marker has shown both positive and negative correlations. Strong staining was found in 9, 14 and 17% of cases of breast cancer in three different studies. Several groups found no relationship between staining and stage, node status, receptor status or size of tumor, but others have shown a significant association between reactivity and mortality or recurrence [13]. One study has demonstrated a correlation between *ERBB2* overexpression and the presence of hematogenous metastases.

These conflicting results are likely to be due to variations in the numbers of patients examined, differences in antibodies used, length of follow-up and methods of scoring immunoreactivity. The immunohistochemical technique has potential as a routine method for the analysis of expression of *ERBB2* in tumors, once all the parameters have been carefully evaluated and standardized. Data obtained from such studies should confirm or refute a role for *ERBB2* as a prognostic indicator in this disease.

8.2.3   p53

Numerous studies have shown the importance of the *p53* tumor suppressor gene in cancer, including breast cancer [14,15 and references therein]. In addition to mutations, a second mechanism is believed to inactivate wild-type *p53*. In some types of breast cancer, the wild-type p53 can lose its tumor-suppressive function by being sequestered in the cytoplasm and prevented from entering the nucleus, its normal site of action [16]. This has been suggested as a possible explanation for the development of breast cancer in cases that do not show any *p53* mutations.

Most breast cancer cases are believed to be sporadic, the results of a spontaneously arising mutation. However, a small proportion of cancers result from an inherited predisposition. Evidence for inherited disease include early age of onset and extensive bilateral breast tumors clustered within families. For example, the Li–Fraumeni syndrome is a familial predisposition to the early childhood development of soft tissue sarcomas as well as early onset breast cancer in parents and relatives [14].

Breast tumors from Li–Fraumeni cancer patients are associated with inherited mutations within exon 7 of the *p53* gene. These mutations are primarily CG to TA transitions at CG dinucleotides [14,17]. Sporadic breast tumors are associated with mutations clustered within exons 5,6,7 and 8. However, most other inherited breast cancer cases are rarely caused by germline point mutations in *p53*. Several studies have shown the possible involvement of a number of other tumor suppressor genes and several oncogenes in the development of the disease [17 and references therein].

Although mutations within *p53* are significantly associated with breast cancer development, several other factors can play a role in the disease process. These include, estrogen receptor status, accumulation of p53 protein and most recently the p53-binding murine double minute-2 (*MDM2*) gene product which has been identified as a negative regulator of p53. Recent studies have indicated that *MDM2* overexpression may be another mechanism by which cancer cells overcome *p53*-regulated growth control without selecting for *p53* mutations *per se* [18].

Familial breast cancer.   Many studies have shown that, with the exception of the Li–Fraumeni syndrome, *p53* gene mutations are not common in breast cancer patients with a family history of malignancy. In a study of five families with a history of breast cancer, none showed any mutations of the *p53* gene [19]. The eight cases that demonstrated a *p53* mutation did not have any family history of the disease. The authors suggested, therefore, that a mutation in another tumor suppressor gene on chromosome 17p is involved in heritable breast cancer.

In a related study, 126 patients with early onset breast cancer were screened and only one of the affected patients was identified with an inherited germ-line *p53* mutation [20]. A study of 25 families with a strong family history of breast cancer demonstrated a complete absence of any germ-line *p53* mutations [14]. An independent study obtained similar findings and these authors concluded that germ-line tumor suppressor *p53* mutations occur in extremely few breast cancer patients that do not have family histories suggestive of the Li–Fraumeni syndrome [21].

Five to ten percent of breast cancers have been suggested to be caused by highly penetrant autosomal dominantly inherited susceptibility genes and around one in 200 women in the general population will develop breast cancer as a result of inheriting one of these genes. There are three main syndromes which show an increased risk of breast cancer: (1) Li–Fraumeni syndrome, associated with mutations in *p53* as discussed above; (2) site-specific breast cancer; and (3) breast–ovarian cancer syndrome. Genes for these last two syndromes have now been identified which means that accurate presymptomatic testing of women can be carried out in 'at risk' families.

Sporadic breast cancer.   Sporadic breast cancer cases tend to show a greater incidence of *p53* mutations. In one study 13% of sporadic breast tumors were shown to exhibit *p53* mutations [22]. However, this analysis only included exons 5 and 6, thus the authors speculated that further studies, especially with exons 7 and 8 would reveal a greater mutation frequency.

A mutation frequency of 46% amongst sporadic breast tumors has been obtained in other studies [17,23]. GC to TA transversions and CG to TA transitions are common in tumors caused by mutagenic factors (such as carcinogens in tobacco), suggesting that some exogenous carcinogens also play a role in breast cancer development [17].

Thus from a comparison of both sporadic and hereditary breast cancer studies, it can be concluded that *p53* mutation is a rare etiological factor in hereditary breast cancer but the mutation rate is significantly higher in sporadic tumors.

Prognostic significance of* p53.   When assessing the relevance of alterations in *p53* to breast cancer, and particularly to prognosis, it is important to recognize the limitations common to many of the studies [18]. The two techniques most often used to study *p53* are immunohistochemical staining, to detect protein accumulation, and nucleic acid sequencing. With regard to immunohistochemical studies there are variations at every step, from patient selection, through specimen processing, actual immunohistochemical technique used, interpretation of the data and statistical analyses [24]. As a consequence of this variation, standardization of the immunohistochemical methods being used, so that a more meaningful analysis of results can be obtained, has been proposed by several groups [18,24,25].

With the nucleic acid approach, many investigations have only concentrated on the four highly conserved domains encompassed by exons 5–8, however, *p53* mutations have been detected outside this region and, thus, the number of *p53* mutations may be underestimated [18].

Another area of concern is the independence of *p53* as a prognostic variable from other better-characterized predictors as well as the striking correlation between the number of cases examined in a particular study and the likelihood of *p53* being found to be a prognostic factor [24]. It should be noted that these limitations also apply to studies of *p53* in other cancers as well as to the study of other oncogenes.

With these limitations in mind, several researchers have demonstrated that the detection of mutant *p53* in patients with breast cancer has significant prognostic value [26–30]. Patients with node negative breast cancer that have overexpressed mutant p53 proteins and/or mutations within the *p53* gene have a much lower chance of long-term survival and a greater incidence of recurrence following surgery. Thus p53 accumulates within the cell nucleus and can be detected by immunohistochemistry [29,30]. Similarly, point mutations within the *p53* gene which can be detected by SSCP analysis are also indicators of poorer prognosis. This is not surprising considering that such a mutation can result in an altered protein conformation resulting in prolongation of its half-life [5, 31].

A recent study in which the entire coding region of *p53* was sequenced in 316 consecutive cases of breast cancer showed that mutations in the evolutionary conserved regions II and IV of the p53 protein (codons 117–142 and 270–286 respectively) were associated with a significantly worse prognosis [32]. In another study of 392 breast carcinomas analyzed immunohistochemically, abnormal expression of p53 protein was found to be an independent prognostic indicator, although it was only a weak one [33].

***Responsiveness to therapy and* p53.**   Using the nuclear accumulation of p53 as a surrogate marker of alteration in *p53* function, no statistically significant evidence was found that p53 status, as assessed by immunohistochemistry, can predict the response to combination chemotherapy in lymph node-negative breast cancer patients. There was, however, a trend for patients who were p53 negative by immunohistochemistry to receive greater benefit from chemotherapy, but the sample size was too small for this observation to be statistically meaningful [34]. In another study, therapy, especially adjuvant tamoxifen, was found to be of less value to patients with *p53* mutations and lymph node-positive tumors [32].

A comparison of four anti-p53 antibodies in an immunohistochemical analysis of 245 breast cancers showed a significant relationship between immunodetection of p53 with each of the four antibodies and poor responsiveness to endocrine therapy [35]. However, as pointed out by these authors, the evaluation of immunohistochemical detection of p53 protein should be carried out with great care as the observed tumor phenotypes do not always reflect the underlying biochemical and biological cellular changes and hence it is no longer sufficient to designate cases as 'p53 positive'.

The use of adjuvant chemotherapy and toxic drugs for women with node-negative breast tumors has been hotly debated by clinicians. Since node-negative tumors have a much lower malignant potential (than axillary lymph node-positive tumors), post-surgical therapy along with its associated risks are not always necessary. Still 20–30% of these cancer patients will relapse [27]. It is possible that *p53* status may provide a mechanism for screening patients at greater risk of relapse or recurrence of disease. Thus more aggressive post-surgical therapy would be warranted for those women showing overexpression of mutant p53 protein and/or *p53* gene mutations and this is providing a major impetus for clarification of the role of *p53* in breast cancer.

8.2.4   BRCA1 *and* BRCA2

In 1990, linkage was established in early onset breast cancer families to the anonymous DNA marker D17S74 estimated to lie close to a breast cancer susceptibility gene, termed *BRCA1*, on chromosome 17q [36]. Analysis of over 200 families by the Breast Cancer Linkage Consortium indicated that this gene was likely to be involved in the majority of families with breast–ovarian cancer though in only around 50% of families with breast cancer only [37] implicating the existence of more than one gene causing the same disorder. A set of highly informative markers were identified which could be used, both to try to narrow down the region in which *BRCA1* lay and hence enable its isolation (see Chapter 2), as well as to allow linkage analysis to be carried out in families to predict those at high risk of developing the disease.

These linkage studies required considerable care due to the existence of genetic heterogeneity and meant that analysis could only be carried out in families large enough to confirm that the disease was caused by the susceptibility locus on chromosome 17q. In the UK, the Cancer Family Study Group suggested that testing should be restricted to families in whom: (1) at least four members had either breast or ovarian cancers; (2) at least two of the cancers were diagnosed before age 50; (3) the probability of linkage to the *BRCA1* locus was in excess of 95%. A relatively small number of families therefore underwent predictive testing in the early 1990s, an example of which is shown in *Figure 8.1*.

Following the isolation of *BRCA1* in 1994, presymptomatic testing can now be offered to a wider group of at-risk women although there remain some important counselling issues, such as the absence of a completely effective screening protocol and the incomplete penetrance of the gene. For example, several studies have now confirmed that the risk of developing breast cancer, if carrying a mutation in *BRCA1*, is about 70–90% by age 70 and not 100%.

Mutation detection is also a complicated procedure. Over 100 different mutations have been identified so far which are widely spread throughout the

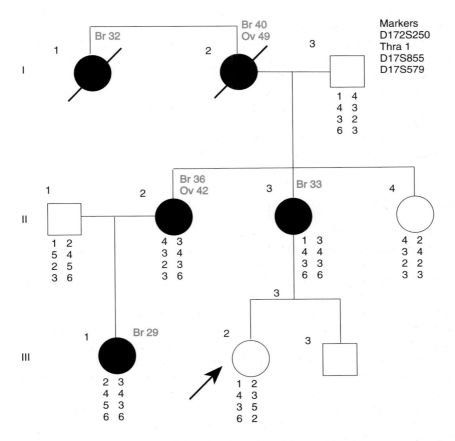

Figure 8.1: Presymptomatic testing was requested by individual III-2 because of a family history of breast and ovarian cancer. The family was tested with four markers linked to *BRCA1* and the disease shown to segregate with haplotype 3,4,3,6. Individual III-2 has therefore inherited the low risk haplotype from her mother and will therefore be at low risk of developing hereditary breast cancer. The details above individuals I-1, I-2, II-2, II-3 and III-1 indicate the ages at which breast and ovarian cancer were diagnosed.

gene [36,38]. Over 90% of the mutations are frameshift, nonsense or splice site mutations leading to a predicted truncated protein product. This is a useful finding as it means that the protein truncation test (PTT) (see Chapter 13) can be used to identify the majority of mutations (*Figure 8.2*). Around 5% of mutations are missense mutations resulting in the loss of conserved cysteine residues in the RING finger motif and are therefore likely to be genuine causative mutations. It remains to be determined whether other missense mutations are truly disease causing. Although the diversity of *BRCA1* mutations is considerable, approximately 30% of mutations have been seen more than once [39]. The most striking example is the 2 bp deletion at nucleotide 185 (185delAG) which has been found in 1 in 100 Ashkenazi Jews. This mutation, together with a C insertion at nucleotide 5382, has been identified in at least 10% of families in whom the *BRCA1* gene has been completely sequenced [39]. This particular result appears to be specific to North American families as in a recent study, of primarily UK families, each mutation was detected only

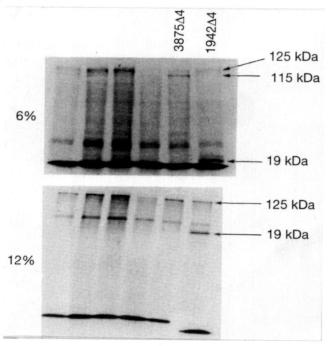

Figure 8.2:   Protein truncation test for exon 11 of *BRCA1*. The individual in lane 5 in the upper gel shows a truncated protein product of 115 kDa compared with the normal band at 125 kDa indicating the presence of a mutation. The individual in lane 6 has a truncating mutation giving a product of 19 kDa which is more clearly seen on the lower 12% gel than the upper 6% gel. Photograph courtesy of Ann Dalton, North Trent Molecular Genetics Service, Sheffield.

once [40]. However, a number of other mutations have been identified relatively frequently in studies world-wide.

A preliminary correlation has been made between genotype and phenotype [40]. There is a significant correlation between the ratio of breast to ovarian cancers in families and the location of the mutation, such that mutations in the 3′ third of the gene are associated with a lower incidence of ovarian cancer. The transition point for this change in ovarian cancer risk appears to be around exon 13 but larger numbers are still required to confirm this finding.

Presymptomatic testing in families with proven *BRCA1* mutations is now underway at many centers. However, the diversity of mutations, plus the fact that *BRCA1* has not yet been shown to have any role in sporadic breast cancer, means that population screening is not yet possible.

The gene for *BRCA2* has now also been isolated and a small number of mutations so far identified. Mutations in *BRCA2* confer a similar risk for the development of breast cancer as do mutations in *BRCA1*, but families show a lower incidence of ovarian cancer. In addition, *BRCA2* mutations account for as much as 15% of all male breast cancer cases. Preliminary evidence suggests that this gene may also confer an increased risk of prostate, laryngeal and endometrial cancer [38]. More extensive mutation analysis of this gene should soon be possible.

Ataxia telangiectasia (AT) is an extremely rare condition with an incidence for homozygotes of around 1 in 100 000 births. It is associated with progressive neuromotor degeneration and the appearance of dilated blood vessels (telangiectases) in the eye. Approximately 1% of the population are heterozygous for AT. These heterozygotes have been shown to have an increased cancer predisposition three- to four-fold that of the general population. In particular, female heterozygotes are at a five-fold increased risk of breast cancer compared to normal women and it has been suggested that AT carriers could account for up to 7% of all breast cancer cases. The gene for this disorder has recently been cloned and therefore opens up new opportunities to study its role in breast cancer predisposition in the general population [41].

8.2.5   CCND1

Amplification of the 11q13 region on chromosome 11 is associated with a number of cancers, including breast carcinoma. The gene for cyclin D1 (*CCND1/PRAD1*) has been located in this region, near to the *BCL1* breakpoint, and its association with a number of cancers, especially B-cell tumors, qualifies it as an oncogene [42]. In a study of 226 breast carcinomas 11q13 amplification was associated with a significantly shorter relapse-free survival and a higher breast cancer-specific mortality suggesting that 11q13 amplification identifies a subgroup of node-positive patients at high risk [43]. At the present time it is still not resolved whether amplification of the 11q13 region reflects a selection for *CCND1*, or other genes which have been also identified there (*HST1, INT2,*

BCL1). As a result, there is no direct evidence that the *CCND1* gene has a causative role in breast cancer [44].

8.2.6. Male breast cancer

Male breast cancer is rare and represents approximately 1% of all cancers in men, causing approximately 0.1% of male cancer deaths per year [45]. In a study of 23 male breast cancers, LOH on chromosome 11q13 was detected in 68% of informative cases [46]. Mutations in the androgen receptor gene, primarily at codons 607 and 608 which are in the second zinc finger of the DNA binding domain of the gene, have been detected [47]. In addition, the risk of breast cancer in males carrying *BRCA2* mutations, though small, is probably greater than in men carrying *BRCA1* mutations [48].

8.2.7 RB1

RB1 has been found to be disrupted by deletion or chromosomal rearrangement in some breast cancer cell lines leading to loss of wild-type RB expression. However, the correlation between LOH at *RB1* and the loss of RB protein is uncertain as the published literature is conflicting on this point [49 and references therein].

8.2.8 Other chromosomal abnormalities and potential tumor suppressors

In addition to the genetic alterations indicated above there are a number of other reported alterations associated with breast cancer. LOH on the long arm of chromosome 6 has been detected in 48% of primary breast cancers, confirming an earlier study indicating that this is a frequent occurence in primary breast carcinomas [50].

LOH for 7q31 may also be an early event in breast cancer development [51]. Chromosome 1 abnormalities have also been reported to be a frequent occurence and rerrangements of 1p correlate with a poor prognosis [52].

Finally, a recent study has shown that the E-cadherin gene, encoding the cell–cell adhesion molecule E-cadherin, was not mutated in 42 cases of infiltrative ductal or medullary breast carcinomas but was mutated in four out of seven infiltrative lobular breast cancers using a PCR/SSCP method. Although this study suffers from the sample size problem alluded to earlier, if confirmed, the results could offer a molecular explanation for the typical scattered tumor cell growth seen in infiltrative lobular breast carcinoma [53].

References

1. Parker, S.L., Tong, T., Bolden, S.B. and Wingo, P.A. (1996) CA – *Cancer J. Clinic.*, **46**, 5.

2. Baum, M., Saunders, C. and Meredith, S. (1994) *Breast Cancer. A Guide for Every Woman*. Oxford University Press, Oxford.

3. Ursin, G., Bernstein, L. and Pike, M.C. (1994) in *Cancer Surveys, Vol. 19: Trends in Cancer Incidence and Mortality*. Imperial Cancer Research Fund (R. Doll, J.F. Fraumeni Jr. and C.S. Muir, Eds). Cold Spring Harbor Laboratory Press, Cold Spring Harbor, NY, p. 241.

4. Ottesen, G.L., Christensen, I.J., Larsen, J.K., Christiansen, J., Hansen, B. and Andersen, J.A. (1995) *Cytometry*, **22**, 168.

5. Runnebaum, I.B., Nagarajan, M., Bowman, M. *et al*. (1991) *Proc. Natl Acad. Sci. USA*, **88**, 10657.

6. Bieche, I. and Lidereau, R. (1995) *Genes Chrom. Cancer*, **14**, 227.

7. Field, J.K. and Spandidos, D.A. (1990) *Anticancer Res*., **10**, 1.

8. Walker, R.A., Senior, P.V., Jones, J.L., Critchley, D.R. and Varley, J.M. (1989) *J. Pathol*., **158**, 97.

9. Berns, E.J.M.M., Klijn, J.G.M., van Putten, W.L.J., van Staveren, I.L., Portengen, H. and Foekens, J.A. (1992) *Cancer Res*., **52**, 1107.

10. Borg, A., Baldertorp, B., Ferno, M., Olsson, H. and Sigurdsson, H. (1992) *Int. J. Cancer*, **51**, 687.

11. Krontiris, T.G., Devlin, B., Karp, D.D., Robert, N.J. and Risch, N. (1993) *N. Engl. J. Med*., **329**, 517.

12. Slamon, D.J., Godolphin, W., Jones, L.A. *et al*. (1989) *Science*, **244**, 707.

13. Walker, R.A., Gullick, W.J. and Varley, J.M. (1989) *Br. J. Cancer*, **60**, 426.

14. Warren, W., Eeles, R.A., Ponder, B.A.J. *et al*. (1992) *Oncogene*, **7**, 1043.

15. Delmolino, L., Band, H. and Band, V. (1993) *Carcinogenesis*, **14**, 827.

16. Moll, U.M., Riou, G. and Levine, A.J. (1992). *Proc. Natl Acad. Sci. USA*, **89**, 7262.

17. Coles, C., Condie, A., Chetty, U. *et al*. (1992) *Cancer Res*., **52**, 5291.

18. Ozbun, M.A. and Butel, J.S. (1995) *Adv. Cancer Res*., **66**, 71.

19. Prosser, J., Porter, D., Coles, L. *et al*. (1992). *Br. J. Cancer*, **65**, 527.

20. Sidransky, D., Tokino, T., Helzlsouer, K.. *et al*. (1992) *Cancer Res*., **52**, 2984.

21. Borresen, A.L., Andersen, T.I., Garber, J. *et al*. (1992). *Cancer Res*., **52**, 3234.

22. Prosser, J., Thompson, A.M., Cranston, G. and Evans, H.J. (1990) *Oncogene*, **5**, 1573.

23. Osbourne, R.J., Merlo, G.R., Mitsudomi, T. *et al*. (1991) *Cancer Res*., **51**, 6194.

24. Dowell, S.P. and Hall, P.A. (1995) *J. Pathol*., **177**, 221.

25. Silvestrini, R., Rao, S., Benini, E., Daidone, M.G. and Pilotti, S. (1995) *J. Natl Cancer Inst*., **87**, 1020.

26. Bosari, S., Lee, A.K.C., Viale, G., Heatley, G.J. and Coggi, G. (1992) *Virch. Archiv. Pathol. Anat*., **421**, 291.

27. Callahan, R. (1992) *J. Natl Cancer Inst*., **84**, 826.

28. Isola, J., Visakorpi, T., Holli, K. and Kallioniemi, O.-P. (1992) *J. Natl Cancer Inst*., **84**, 1109.

29. Thor, A.D., Moore, D.H., Edgerton, S.M. *et al*. (1992) *J. Natl Cancer Inst*., **84**, 845.

30. Allred, D.C., Clark, G.M., Elledge, R. *et al*. (1993) *J. Natl Cancer Inst*., **85**, 200.

31. Elledge, R.M., Fugua, S.A.W., Clark, G.M. *et al*. (1993) *Breast Cancer Res. Treat*., **26**, 225.

32. Bergh, J., Norberg, T., Sjogren, S., Lindgren, A. and Holmberg, L. (1995) *Nature Med*., **1**, 1029.

33. Pietilainen, T., Lipponen, P., Aaltoma, S., Eskelinen, M., Kosma, V.-M. and

Syrjanen, K. (1995) *J. Pathol.*,**177**, 225.

34. Elledge, R.M., Gray, R., Mansour, E. *et al.* (1995) *J. Natl Cancer Inst.*, **87**, 1254.
35. Horne, G.M., Anderson, J.J., Tiniakos, D.G., McIntosh, G.G., Thomas, M.D., Angus, B., Henry, J.A., Lennard, T.W.J. and Horne, C.H.W. (1996). *Br. J. Cancer*, **73**, 29.
36. Szabo, C.I. and King, M.C. (1995) *Hum. Mol. Genet.*, **4**, 1811.
37. Easton, D.F., Bishop, D.T., Ford, D. and Crockford, G.P. (1993) *Am. J. Hum. Genet.*, **52**, 678.
38. Stratton, M.R. and Wooster, R. (1996) *Curr. Opin. Genet. Devel.*, **6**, 93.
39. Shattuck-Eidens, D., McClure, M., Simard, J., Labrie, F., Narod, S. and Couch, F. (1995) *J. Am. Med. Assoc.* **273**, 535.
40. Gayther, S., Warren, W., Mazoyer, S., Russell, P.A., Harrington, P.A., Chiano and M. (1996) *Nature Genetics*, **11**, 428.
41. Savitsky, K., Bar-Shira, A., Giald, S., Rotman, G., Ziv, Y and Vanagaite, L. (1995) *Science*, **268**, 1749.
42. Gillet, C., Fantl, V., Smith, R., Fisher, C., Bartek, J., Barnes, D. and Peters, G. (1994) *Cancer Res.*, **54**, 1812.
43. Schurring, E., Verhoeven, E., van Tinteren, H. *et al.* (1992) *Cancer Res.*, **52**, 5229.
44. Dickson, C., Fantl, V., Gillet, C., Brookes, S., Bartek, J., Smith, R., Fisher, C., Barnes, D. and Peters, G. (1995) *Cancer Lett.*, **90**, 43.
45. Thomas, D.B. (1993) *Epidemiol. Rev.*, **15**, 220.
46. Sanz-Ortega, J., Chuaqui, R., Zhuang, Z., Sobel, M.E., Sanz-Esponera, J., Liotta, L.A., Emmert-Buck, M.R. and Merino, M.J. (1995) *J. Natl Cancer Inst.*, **87**, 1408.
47. Wooster, R., Mangion, J., Eeles, R., Smith, S., Dowsett, M., Averill, D., Barrett-Lee, P., Easton, D.F., Ponder, B.A. and Stratton, M.R. (1992) *Nature Genetics*, **2**, 132.
48. Wooster, R., Neuhausen, S.L., Mangion, J. *et al.* (1994) *Science*, **265**, 2088.
49. Devilee, P. and Cornelisse, C.J. (1994) *Biochim. Biophys. Acta*, **1198**, 113.
50. Zheng, Z.M., Marchetti, A., Buttita, F., Champeme, M-H., Campani, D., Bistochi, M., Lidereau, R. and Callahan, R. (1996) *Br. J. Cancer*, **73**, 144.
51. Champeme, M.H., Bieche, I., Beuzelin, M. and Lidereau, R. (1995) *Genes Chrom. Cancer*, **12**, 304.
52. Hainsworth, P.J., Raphael, K.L., Stillwell, R.G., Bennett, R.C. and Garson, O.M. (1992) *Br. J. Cancer*, **66**, 131.
53. Berx, G., Cleton-Jansen, A.-M., Nollet, F., de Leeuw, W.J.F., van de Vijver, J., Cornelisse, C. and van Roy, F. (1995) *EMBO J.*, **14**, 6107.

Chapter 9

Genitourinary cancer

9.1 Ovarian and endometrial cancers

It is difficult to summarize international trends in the incidence and mortality of ovarian cancer since they differ between countries and between age groups [1]. In recent years the mortality rate has started to decline in the younger age groups (under 55 years) but this has not been reflected in the incidence data.

There are several similarities between ovarian and breast cancers; both often express steroid hormone receptors and it is suggested that they have common etiological factors. Several prognostic factors have been identified in ovarian cancer, including stage of disease, presence or absence of ascites and volume of residual disease following surgery. Histological grading and ploidy are also strongly associated with clinical outcome.

Molecular genetic studies of ovarian cancer have been limited by the inaccessible location of the ovary, the advanced stage of tumors available for analysis, and the lack of a well-defined precursor lesion [2]. Amplification of *MYC* occurs in approximately 30% of ovarian cancers and is more frequently seen in advanced stage serous cancers [3].

In a series of 74 ovarian cancers, *KRAS* mutations were detected in 45% of mucinous, compared to only 16% of serous cancers. In the mucinous tumors, *KRAS* mutations were present in half of the borderline and in an equal number of invasive tumors. *KRAS* mutations were seen in eight of 31 mucinous tumors that lacked allele loss for more than 30 polymorphic markers spanning the human genome. Borderline ovarian tumors constitute a unique subgroup characterized by an unusual degree of epithelial cell proliferation and atypia compared with benign tumors, but they lack the stromal invasion characteristic of invasive ovarian tumors. The clinical course of these tumors is between that of benign and malignant tumors and they metastasize within the peritoneal cavity but rarely result in death. The frequent finding of *KRAS* mutations in borderline mucinous tumors suggests that these mutations are an early event in the development of mucinous cancers of the ovary and that borderline tumors are precursors of invasive tumors [2 and references therein].

Like breast cancers, approximately 30% of ovarian cancers carry amplified *ERBB2*, but here the presence of the amplified oncogene appears to correlate with poor survival. In one study, the survival times for patients with one, two to five, or over 10 copies of *ERBB2* were 1879, 959 and 243 days, respectively. Overexpression of the *HER-2/NEU* oncogene also occurs in approximately 30% of ovarian cancers and increased expression is correlated with poor survival [3 and references therein].

A high frequency of LOH has been detected on chromosomes 6p, 6q, 9q, 13q, 17p and 17q and together with mutation analysis this suggests that chromosome 17 plays a significant role in ovarian cancer development. On the short arm, LOH and mutations of the *p53* locus as well as LOH at a more distal locus (17p13.3) have been observed in a high percentage of tumors. Similarly, on the long arm, losses in the *BRCA1* region and a more distally located locus (17q22-23) are frequently seen [2 and references therein].

It has been estimated that 5–10% of ovarian cancers are the result of an inherited predisposition. The susceptibility to ovarian cancer is inherited through one of three distinct patterns: (1) the breast–ovarian cancer syndrome, which is typified by multiple cases of early age of onset breast and ovarian cancer; (2) hereditary nonpolyposis colorectal cancer, in which excess cancer cases are primarily cancers of the proximal colon, endometrium and ovary; (3) site-specific ovarian cancer, which is the least common. As with retinoblastoma and other inherited cancers, the susceptibility to cancer in families with a history of site-specific ovarian cancer is transmitted as an autosomal dominant trait with an early age of onset [4]. In families with a history of breast–ovarian cancer, the cumulative risk for breast cancer or ovarian cancer in women with a mutant *BRCA1* allele is estimated to be 76% by age 70, thus indicating that *BRCA1* is a highly penetrating predisposing gene for both malignancies [2,3]. It has been suggested that there are two classes of mutation, one which confers a breast cancer risk of 71% and an ovarian cancer risk of 87%, and another which confers a breast cancer risk of 86% and an ovarian cancer risk of only 18%. The lifetime risk of breast cancer was independent of mutation but this was not true for ovarian cancer [5]. Recently it has been shown that it is mutations at the 5' end of the gene which confer the risk of ovarian cancer and that mutations at the 3' end result in minimal risk [6].

Mutation of *p53* with resultant overexpression of mutant p53 protein occurs in 50% of stage III/IV, 15% of stage I/II and 4% of borderline ovarian cancers. Most mutations are transitions, which suggests that they arise spontaneously rather than being due to exogenous carcinogens [3]. Mutations of *p53* in ovarian cancer have been shown to be associated with advanced stage of disease [7,8], poorly differentiated carcinomas [9] and a poor prognosis [10]. In addition, recent results have shown a significant correlation between p53 accumulation, type of mutation (missense) and pathological response to cisplatin-based chemotherapy [11] and that transfection of a wild-type *p53* construct into an ovarian cancer cell line with a mutant *p53* gene can inhibit proliferation

[12]. A recent study of 19 ovarian carcinomas and 17 borderline tumors has confirmed that loss or inactivation of tumor suppressor gene function, by chromosome 17p allelic deletions, or *p53* mutations, are important genetic changes in ovarian cancer [13]. However, the prognostic significance of these observations remains unclear at present.

Interpretation of international trends in the incidence and mortality of endometrial cancers is complicated by variation over time and, in different countries, of the coding of such tumors [1]. Despite this caveat there is an international downward trend in mortality.

Endometrial cancers can be divided into endometrioid adenocarcinomas, which account for the majority of endometrial cancers and are typified by developing from atypical endometrial hyperplasia in the setting of excess estrogenic stimulation. In contrast, serous carcinomas are representative of endometrial cancers in older women who have endometrial atrophy and lack the typical endometrial cancer risk factors reflecting unopposed estrogen exposure [14].

Overexpression of *HER-2/NEU* occurs in 10% of endometrial adenocarcinomas and correlates with intraperitoneal spread of disease and poor survival. *MYC* is amplified in 10% of cases. Point mutations in codon 12 of *KRAS* have been reported to occur in 10–20% of cases. *KRAS* mutations have also been reported in some endometrial hyperplasias, which may represent an early event in the development of some endometrial cancers. Mutations of *p53* occur in 20% of endometrial adenocarcinomas and overexpression of mutant p53 protein is associated with advanced stage and poor survival. Since *p53* mutations have not been seen in endometrial hyperplasias, this is thought to be a relatively late event in endometrial carcinogenesis. However, serous carcinomas are frequently associated with *p53* abnormalities and appear to develop from a surface lesion termed endometrial intraepithelial carcinoma. Although serous carcinomas are rare, these highly aggressive tumors account for a disproportionate number of endometrial cancer deaths [14].

Microsatellite instability has also been noted in approximately 15% of sporadic endometrial cancers, but mutations in DNA repair genes have not yet been reported [15]. Finally, in a recent study of the putative tumor suppressor *E-cadherin* gene in 135 carcinomas of the endometrium and ovary only four mutations were detected [16]. The prognostic significance, if any, of this gene in ovarian and endometrial cancers remains to be determined.

9.2 Cervical cancer

Cancer of the uterine cervix is the second most common cancer in women worldwide and is particularly common in less developed countries, where 80% of the world's cervical cancer occurs. Current evidence indicates that the main cause of cervical cancer is cervical infection with certain types of HPV, particularly types 16, 18, 31, 33, 35 and 45, that are transmitted sexually [17]. The overall incidence and mortality from cervical cancer has declined in western countries and in most developing countries. In women under 40 years of age, however, mortality rates are leveling off or increasing in most countries [17]. The reasons for the overall decline in cervical cancer may be linked to improvements in the general standard of living and the increase in younger women may be due to the increasing prevalence of HPV infection. Screening for cervical cancer has undoubtedly led to a decline in incidence and mortality in many countries.

A number of prognostic factors have been linked to the progression of cervical cancers, including depth of stromal invasion, lesion depth and nodal involvement. In particular, HPV has been found to be integrated into malignant cells and the presence of the virus (in specific subtypes) in premalignant lesions indicates a poor prognosis. These anogenital-associated HPVs are further classified into low risk (e.g. HPV 6, HPV 11), associated with benign genital warts, and high risk (e.g. HPV 16, HPV 18), associated with lesions such as squamous intraepithelial neoplasia (SIN) which can progress to anogenital cancer. Approximately 85% of all cervical carcinomas are conservatively estimated to be high-risk HPV-positive. Epidemiological studies have shown that infection with a high-risk HPV is a significant risk factor for developing cervical cancer [18]. The two HPV proteins E6 and E7 are consistently expressed in cervical carcinomas and are thought to functionally abrogate critical cell cycle regulatory pathways, including those governed by the p53 and RB proteins. Subversion of these pathways by viral proteins causes genomic instability, resulting in the accumulation of chromosomal abnormalities followed by clonal expansion of malignant cells [18].

Abnormal expression of several oncogenes also has prognostic value. Amplification and overexpression of *MYC* has been shown to be more frequent in advanced tumors of the uterine cervix compared with early tumors. In tumor samples from 72 untreated patients with carcinoma of the cervix, overexpression was detected in 35% of cases. This overexpression was not a consequence of amplification which was seen in only 8% of cases. There was no relationship between *MYC* overexpression and stage, nodal status or age. However, there was an eight-fold greater risk of relapse for patients with overexpression of *MYC*, which outweighed even nodal status as a prognostic factor [19].

Elevated expression of RAS p21, as demonstrated by the use of antibody Y13 259, has been found in malignant as opposed to benign or premalignant lesions. In the small-cell type of squamous cell carcinomas, tumors with elevated expression of RAS p21 were shown to have a better prognosis than negative tumors. Expression of RAS p21 together with histological type may therefore be of prognostic significance in carcinomas of the cervix in specific histological types [19]. Mutations at codon 12 of the *HRAS* gene have also been found in cervical cancer and in one study were correlated with poor prognosis [19].

9.3   Testicular cancer

Testicular cancer is a rare disease and unlike most other solid tumors, the incidence rate does not increase as a function of age but reaches a maximum at about 25–34 years and then declines to become uncommon in men over the age of 55 years [20]. Although the incidence has been increasing over the years introduction of effective chemotherapy in the mid 1970s has resulted in reduced mortality rates.

Strong expression of p62 has been detected in differentiating areas of testicular tumors. In a flow-cytometric study of p62 expression in nuclei extracted from paraffin blocks, increasing levels of p62 correlated with increasing differentiation of teratomas. Patients who had no recurrence 3 years after diagnosis showed a significantly higher level of p62 than those who developed a recurrence in this time period. Expression of *MYC* may therefore prove to be of prognostic significance in this disease [21].

9.4   Renal cell cancer and bladder cancer

Although the incidence of bladder cancer is rising in most populations its mortality has begun to fall in a number of countries including the USA and Japan but not, on the whole, in continental Europe. Renal cancer is common in societies with high rates of bladder cancer, suggesting that these two functionally and anatomically related sites may be exposed to similar carcinogenic influences [22]. Both the incidence and mortality of renal parenchymal cancer have been steadily increasing in virtually every country.

The clinical behavior of renal cancer is unpredictable, although a few features, such as stage of tumor and ploidy, are partially correlated with prognosis. The disease is highly metastatic, with 30% of patients having metastases at the time of diagnosis. A study of *LMYC* RFLPs in patients with renal cell cancer has shown no significant differences between the RFLP patterns of normal and tumor DNA. However, the presence of metastases was associated with lack of a particular allele suggesting that this RFLP may be a marker of genetic predisposition to the formation of metastases rather than the development of primary disease.

Von Hippel–Lindau (VHL) disease is an autosomal dominant disorder and, eventually, the majority of patients with this disease will develop a renal cell carcinoma if they live long enough. Renal cell carcinoma is now the leading cause of death in patients with VHL disease [23]. The VHL disease gene has been mapped to the short arm of chromosome 3 and subsequently isolated. Flanking markers have also been identified which can be used in gene tracking studies [23,24 and references therein]. Statistical and molecular genetic studies suggest that the *VHL* gene functions as a tumor suppressor gene like *RB1*. Although chromosome 3p loci are involved in the pathogenesis of sporadic renal cell carcinomas, it is not yet known whether acquired mutations at the

VHL locus are responsible for sporadic renal cell carcinomas [23,24]. VHL disease is the most frequent cause of inherited renal cell carcinoma but familial renal cell carcinomas, without additional features, also occur and are associated with a t(3;8) translocation. In a kindred study renal cell carcinoma was only seen in translocation carriers, each of which had an 87% risk of developing the cancer by 60 years of age [25 and references therein].

The entire urinary tract is lined with transitional cell epithelium and a multistep model for the development of transitional cell carcinoma (TCC) of the bladder, based on experiments in rodents, has been proposed [26] (*Figure 9.1*). EGF in rat urine has been found to be responsible for the potentiating effect of chemical carcinogens on the pathogenesis of TCC. Overexpression of EGFR protein is related to higher cytological grade and more aggressive disease [26]. In a prospective study of 101 new cases of TCC of the bladder, by immunohistochemical staining for EGFR, death was independently associated with advanced stage and EGFR-positivity. In addition, those patients with EGFR-positive tumors were found to have a shorter time to recurrence and a higher recurrence rate [27].

Mutations and LOH of the *RB1* gene are also found in a significant proportion of bladder cancers and are associated with local muscular invasion and a high cytological grading, both of which are related to poor prognosis [26 and references therein]. Overexpression of p53 protein as detected by immunohistochemistry appears to identify those patients with superficial bladder cancer at risk of the development of muscle invasive or metastatic disease. However, the role of p53 overexpression in patients with advanced or metasta-

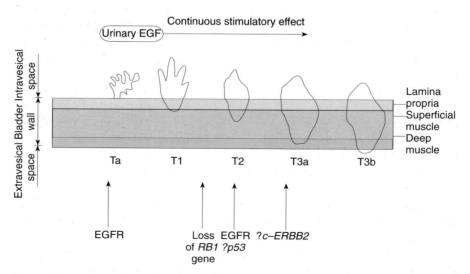

Figure 9.1:   A sketch relating the progression of bladder TCC to known genetic results involved. Reproduced from ref. 26 with permission from Blackwell Science Ltd.

tic bladder cancer is not yet well established. In one study of specimens from 44 patients with advanced bladder tumors (T2–T4) undergoing radical cystectomy, p53 overexpression was found to be a significant independent prognostic indicator. While this needs to be confirmed in prospective studies it provides a potential prognostic tool for decision making for defined subgroups of patients in the future [28].

9.5 Prostatic cancer

Prostate cancer is a common malignancy in men and the incidence rate and mortality are increasing with time in all populations that have been studied [29]. This disease also has an extremely variable natural history, ranging from a noninvasive to a rapidly metastatic disease which is fatal within a short time following diagnosis. It is frequently diagnosed incidentally at the time of surgery during transurethral resection for urinary obstruction. Although factors such as histological grade have been evaluated as prognostic factors, there is considerable uncertainty concerning their use. Tumor markers such as carcinoembryonic antigen and prostatic acid phosphatase have not proved to be consistently useful.

Very little is known about the molecular and genetic mechanisms involved in prostate cancer. Previous studies have shown frequent LOH (40%) at chromosomal regions 8p, 10q, and 16q suggesting the presence of tumor suppressor genes in these regions [30]. Frequent LOH at 7q31.1 in primary prostatic cancer has been shown to be associated with tumor aggressiveness and progression [31]. The recent finding of LOH of the *BRCA1* and other loci on chromosome 17q21 suggests a possible role for genes in this region in the pathogenesis of prostate cancer [32]. Alterations of *p53* [33], and loss of *DCC* expression (see Section 6.2.5) and LOH at the *DCC* locus [34] have also been reported, suggesting a possible role for these genes in the development and/or progression of the disease [30].

Levels of *MYC* have been determined in prostate cancer and benign prostatic hyperplasia to determine whether the level of expression of the gene can distinguish between malignant and benign disease or can predict aggressive disease. Significantly higher levels of *MYC* expression have been found in malignant disease and a subset of patients had increased levels compared with the rest of the cancer group. At present the prognostic value of these findings remains to be determined.

Expression of RAS p21 has been examined in prostate cancer using RAP 5 antibody. Staining was particularly associated with high grade, and presumably more metastatic, tumors. In a semi-quantitative assay, there was a significant association between degree of nuclear aplasia and expression of RAS p21. Compared with the tumor markers carcinoembryonic antigen and prostate-specific antigen (PSA), RAS p21 was the only phenotypic marker which correlated with tumor grade.

Recently a metastasis suppressor gene for prostate cancer has been identified on human chromosome 11p11.2, designated *KAI1*. Expression of *KAI1* was reduced in human cell lines derived from metastatic prostate cancers. *KAI1* specifies a protein of 267 amino acids, with four hydrophobic (presumably transmembrane) domains and one large extracellular domain with three potential glycosylation sites. The gene is evolutionarily conserved, expressed in many human tissues, and encodes a member of a structurally distinct family of leukocyte surface glycoproteins [35].

Recent work with the LNCaP cell line, established from a metastatic lesion of human prostate adenocarcinoma which carries a t(6;16)(p21;q22) translocation has shown that the breakpoint is within the haptoglobin gene cluster on chromosome 16 and within a novel gene, *TPC* (translocated in prostate cancer) similar to the prokaryotic S10 ribosomal protein gene, on chromosome 6. The translocation results in the production of a fusion transcript, TPC/HPR (haptoglobin related) which may provide important clues in the future about the pathogenesis of prostate cancer [30].

One other gene that appears promising for prostate cancer is *E-cadherin*, coding for a surface adhesion molecule, which, like KAI1, is a membrane protein involved in cell–cell interactions. E-cadherin was reduced in advanced stage tumors in 90 prostate samples and in a subsequent 3 year follow-up study, men whose tumors showed normal amounts of E-cadherin showed no disease progression while those whose tumors showed little expression of the protein had high PSA levels, and their disease had advanced [36].

References

1. Mant, J.W.F. and Vessey, M.P. (1994) in *Cancer Surveys*, Vol. 19: *Trends in Cancer Incidence and Mortality*. Imperial Cancer Research Fund (R. Doll, J.F. Fraumeni Jr. and C.S. Muir, Eds). Cold Spring Harbor Laboratory Press, Cold Spring Harbor, NY p. 175.
2. Gallion, H.M., Pieretti, M., DePriest, P. D. and van Nagell, J.R., Jr. (1995) *Cancer*, **76**, 1992.
3. Berchuck, A. (1995) *J. Cell. Biochem.* **23** (Suppl), 223.
4. Gallion, H.H. and Smith, S.A. (1994) *Semin. Surg. Oncol.*, **10**, 249.
5. Ford, D., Easton, D., Bishop, D.T., Narod, S.A., Goldgar, D.E. (1994) *Lancet*, **344**, 761.
6. Gayther, S., Warren, W., Mazoyer, S. *et al.* (1996) *Nature Genetics*, **11**, 428.
7. Eccles, D.M., Brett, L., Lessells, A., Gruber, L., Lane, D., Steel, C.M. and Leonard, R.C.F. (1992) *Br. J. Cancer*, **65**, 40.
8. Sheridan, E., Hancock, B.W. and Goyns, M.H. (1993) *Cancer Lett.*, **68**, 83.
9. Kihana, T., Tsuda, H., Teshima, S., Okada, S., Matsuura, S. and Hirohashi, S. (1992) *Jpn. J. Cancer Res.*, **83**, 978.
10. Bosari, S., Viale, G., Radaelli, U., Bossi, P., Bonoldi, E. and Coggi, G. (1993) *Human Pathol.*, **24**, 1175.
11. Righetti, S.C., Torre, G.D., Pilotti, S. *et al.* (1996) *Cancer Res.*, **56**, 689.
12. Santoso, J.T., Tang, D.-C., Lane, S.B. *et al.* (1995) *Gynecologic Oncol.*, **59**, 171.

13. Lee, J.-H., Kang, Y.-S., Park, S.-Y., Kim, B.-G., Lee, E.-D., Lee, K.-H., Park, K.-B., Kavanagh, J.J. and Wharton, J.T. (1995) *Cancer Genet. Cytogenet.*, **85**, 43.
14. Sherman, M.E., Sturgeon, S., Brinton, L. and Kurman, R.J. (1995) *J. Cell. Biochem.* **23** (Suppl), 160.
15. Berchuck, A. (1995) *J. Cell. Biochem.* **23** (Suppl), 174.
16. Risinger, J.I., Berchuck, A., Kohler, M.F. and Boyd, J. (1994) *Nature Genetics*, **7**, 98.
17. Beral, V., Hermon, C., Munoz, N. and Devesa, S.S. (1994) in C*ancer Surveys*, Vol. 19; *Trends in Cancer Incidence and Mortality*. Imperial Cancer Research Fund (R. Doll, J.F. Fraumeni Jr and C.S. Muir, Eds). Cold Spring Harbor Laboratory Press, Cold Spring Harbor, NY, p. 265.
18. Munger, K. (1995) *J.Cell. Biochem.* **23** (Suppl), 55 (and references therein).
19. Field, J.K. and Spandidos, D.A. (1990) *Anticancer Res.*, **10**, 1.
20. Forman, D. and Moller, H. (1994) in *Cancer Surveys*, Vol. 19: *Trends in Cancer Incidence and Mortality*. Imperial Cancer Research Fund (R. Doll, J.F. Fraumeni Jr and C.S. Muir, Eds). Cold Spring Harbor Laboratory Press, Cold Spring Harbor, NY, p. 175.
21. Watson, J.V., Stewart, J., Evan, G., Ritson, A. and Sikora, K. (1986) *Br. J. Cancer*, **53**, 331.
22. McCredie, M. (1994) in *Cancer Surveys*, Vol. 19: *Trends in Cancer Incidence and Mortality*. Imperial Cancer Research Fund (R. Doll, J.F. Fraumeni Jr and C.S. Muir, Eds). Cold Spring Harbor Laboratory Press, Cold Spring Harbor, NY, p. 343.
23. Hodgson, S.V. and Maher, E.R. (1993) in *A Practical Guide to Human Cancer Genetics*. Cambridge University Press, Cambridge, p. 157.
24. Gnarra, J.R., Duan, D.R., Weng, Y. *et al.* (1996) *Biochim. Biophys. Acta*, **1242**, 201.
25. Hodgson, S.V. and Maher, E.R. (1993) in *A Practical Guide to Human Cancer Genetics*. Cambridge University Press, Cambridge, p. 74.
26. Leung, H.Y. (1994) in *Cancer. A Molecular Approach* (N. Lemoine, J. Neoptolemos, and T. Cooke, Eds). Blackwell Scientific Press, Oxford, p. 223.
27. Neal, D.E., Sharples, L., Smith, K., Fennelly, J., Hall, R.R. and Harris, A.L. (1990) *Cancer*, **65**, 1619.
28. Kuczyk, M.A., Bokemeyer, C., Serth, J., Hervatin, C., Oelke, M., Hofner, K., Tan, H.K. and Jonas, U. (1995) *Eur. J. Cancer*, **31A**, 2243.
29. Whittemore, A.S. (1994) in *Cancer Surveys*, Vol. 19; *Trends in Cancer Incidence and Mortality*. Imperial Cancer Research Fund (R. Doll, J.F. Fraumeni Jr and C.S. Muir, Eds). Cold Spring Harbor Laboratory Press, Cold Spring Harbor, NY, p. 309.
30. Veronese, M.L., Bullrich, F., Negrini, M. and Croce, C.M. (1996) *Cancer Res.*, **56**, 728.
31. Takahashi, S., Shan, A.L., Ritland, S.R. *et al.* (1995) *Cancer Res.*, **55**, 4114.
32. Gao, X., Zacharek, A., Salkowski, A., Grignon, D.J., Sakr, W., Porter, A.T. and Honn, K.V. (1995) *Cancer Res.*, **55**, 1002.
33. Kubota, Y., Shuin, T., Uemura, H. *et al.* (1995) *Prostate*, **27**, 18.
34. Gao, X., Honn, K.V., Grignon, D., Sakr, W. and Chen, Y.Q. (1993) *Cancer Res.*, **53**, 2723.
35. Dong, J-T., Lamb, P.W., Rinker-Schaeffer, C.W., Vukanovic, J., Ichikawa, T., Isaacs, J.T. and Barrett, J.C. (1995) *Science*, **268**, 884.
36. Nelson, N.J. (1995). *J. Natl Cancer Inst.*, **87**, 1281.

Chapter 10

Hematological malignancies

10.1 Introduction

The age-adjusted death rates for hematological malignancies in the major industrialized countries are shown in *Table 10.1*.

Table 10.1: Age-adjusted death rates in the major industrialized countries of the world

Country	Age-adjusted death rates (1990–1993) per 100 000 population	
	Male	Female
Italy[a]	6.5	4.0
USA[b]	6.4	3.9
Canada[b]	6.0	3.7
France[b]	5.9	3.6
Germany	5.9	3.6
Japan	4.2.	2.7
UK[c]	5.1	3.1

Abstracted from ref. 1.
[a]1990–1991; [b]1990–1992; [c]1992–1993.

Many oncogenes have been identified in leukemias and lymphomas because of the presence of consistent chromosomal translocations which can be identified cytogenetically. However, there is relatively little evidence that the wide variety of genes now known to be associated with chromosome translocations are actually capable of causing tumors *in vivo*. It is assumed from their consistent association with translocations that they are oncogenes, but functional evidence is often lacking [2]. However, the breakpoint in one of the chromosomes has frequently been shown to occur in, or close to, an oncogene (*Table 10.2*).

Table 10.2: Chromosomal translocations in leukemias and lymphomas associated with oncogenes

Disease	Translocation	Oncogene
B-CLL	t(11;14)(q13;q32.3) t(14;19)(q32;q13)	*BCL1 (CCND1/cyclin D1)* *BCL3*
Burkitt's lymphoma	t(8;14)(q24.1;q32.3) t(8;22)(q24.1;q11) t(2;8)(p12;q24.1)	*MYC*
Follicular lymphoma	t(14;18)(q32.3;q21.3)	*BCL2*
CML (AML and ALL)	t(9;22)(q34;q11)	*ABL/BCR*
ALL	t(4;11)(q21;q23)	*ALL1*
T-ALL	t(11;14)(p15;q11) t(11;14)(p13;q11) t(2;8)(q24;q24) t(8;14)(q24;q11)	*RBTN1 (TTG1)* *RBTN2 (TTG2)* *MYC* *MYC*

The numerous chromosomal alterations and chromosomal translocations that occur in hematological malignancies are thought to cause oncogenic activation through one of two mechanisms. The first is gene activation after translocations involving the T-cell receptor (TCR) or immunoglobulin gene loci and the second is gene fusion, which occurs when the TCR or immunoglobulin genes are not involved in the translocation process and results in the fusion of the coding regions of two genes which in turn leads to a fusion or chimeric protein with altered functional properties [2,3 and references therein]. Several fusion proteins generated by chromosomal translocations are chimeric transcription factors, that is, they are new transcription factors formed by the fusion of portions of two transcription factors. The activity of these fusion proteins differs considerably from their normal counterparts and they are thought to deregulate the expression of target genes controlled by the wild-type transcription factors. An example of this is the BCR–ABL fusion protein (see Section 10.5).

A 'master gene' model has been postulated to explain the consistent observation in acute leukemias that genes involved in transcriptional regulation are activated by association with TCR or immunoglobulin genes after chromosomal translocation. These activation processes cause the transcription of the master genes in cells that would not normally express them. In cancers, the master gene products bind to their specific activation targets and cause responder target genes to be transcribed. These in turn may bind to specific targets in the genome and activate transcription [2,4]. By acting positively to up-regulate critical target genes, or negatively to interfere with normal regulatory pathways, such transcription factors can disrupt gene regulatory cascades that coordinate the expression of large numbers of proteins required for the completion of specific cell differentiation programs [4].

In addition, other oncogenes commonly identified in human tumors, such as the *RAS* family, have also been found to be associated with hematological malignancies.

10.2 *MYC*

The involvement of the *MYC* gene in Burkitt's lymphoma is probably one of the most studied of all events involving oncogenes. The translocation between chromosomes 8 and 14 is a highly consistent and specific rearrangement found in 75–85% of Burkitt's lymphoma patients. In the remaining 15–25% of patients the translocation occurs between chromosomes 8 and either 2 or 22. This translocation places the *MYC* gene from chromosome 8 next to the immunoglobulin heavy chain locus on chromosome 14 or the κ and λ constant region exons, on chromosomes 2 and 22, 3′ of the *MYC* gene on chromosome 8 (*Figures 10.1* and *2.12*). In the commonest translocation the breakpoint on chromosome 8 occurs on the 5′ side of the second exon of the *MYC* gene. This means that the two coding exons of the gene, exons 2 and 3, are always translocated and placed close to the constant region of the immunoglobulin heavy chain locus on chromosome 14. In the two less common translocations the breakpoint is distal to the *MYC* gene. In all cases, the coding exons of *MYC* do not undergo any structural rearrangement. Instead the translocation results in deregulation of *MYC*. Following the general rule that the development of malignancy is a multistage event, other changes are necessary for the development of Burkitt's lymphoma. In one model the first change is chronic stimulation and activation of the B cells by infectious agents, primarily Epstein–Barr virus (EBV), leading to the emergence of immortalized clones. The second stage is the translocation involving *MYC*. A second model proposes that these two events occur in the reverse order. Additional changes in other oncogenes are likely to follow these two steps [5].

A study of increased *MYC* expression in other malignant lymphomas, using the antibody Myc1–6E10, did not show any significant association, although T-cell immunoblastic malignant lymphomas were consistently positively stained. Flow-cytometric detection of p62 has demonstrated a correlation between *MYC* protein levels and the aggressiveness of malignant non-Hodgkin's lymphomas of B-cell origin (B-NHLs).

Preliminary evidence has suggested that there is variation in an RFLP associated with the *LMYC* locus, which plays a role in both survival and susceptibility to B-NHL and acute lymphoblastic lymphoma.

10.3 *RAS*

Mutations in *RAS* genes, particularly *NRAS*, occur frequently in acute myeloid leukemia (AML) and less commonly in chronic myeloid leukemia (CML) [6]. In myelodysplastic syndrome (MDS) and AML, a mutation has

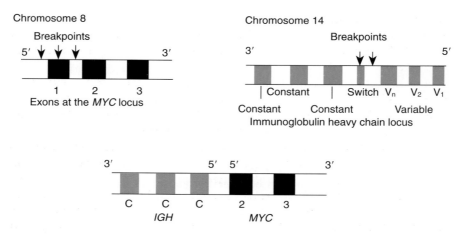

Figure 10.1: Molecular basis of the translocation between the *MYC* gene on chromosome 8 and the immunoglobulin heavy chain on chromosome 14 which occurs in the majority of cases of Burkitt's lymphoma.

been found in the *RAS* gene in 20–30% of cases: this is usually in *NRAS*, occasionally in *KRAS* and infrequently in *HRAS*. Several studies have suggested that patients with MDS and a *RAS* mutation are more likely to progress to AML and hence have the worst prognosis. However, as there are examples of MDS patients with a *RAS* mutation in their multipotent stem cells who have stable disease, this marker is unlikely to be of use on its own as a prognostic indicator. Secondly, AML patients with or without *RAS* mutations do not have a significantly different prognosis and no relationship with degree of differentiation can be seen [6]. Alternative roles for the *RAS* mutation may be as a marker for monitoring the effects of chemotherapy or for detecting minimal residual disease [7].

In a large study of childhood acute lymphocytic leukemia (ALL), point mutations were found in 6% of cases at codons 12 or 13 of *NRAS*. These cases had a significantly higher risk of hematological relapse and a trend towards a lower rate of complete remission than those without *RAS* mutations. Presence of *RAS* mutations was independent of other high-risk factors for ALL and may therefore be a useful prognostic indicator.

10.4   *BCL1* and *BCL2*

In some B-cell lymphoproliferative diseases, including 30–39% of mantle cell lymphomas (MCLs), a t(11;14)(q13;q32) translocation has been observed [8,9]. This translocation involves the immunoglobulin heavy chain locus at chromosome 14q32 and the chromosome 11q13 region and is analogous to the t(8;14) translocation described earlier where *MYC* is placed adjacent to the immunoglobulin heavy chain enhancer. The translocated 11q13 region is known to involve the *BCL1* oncogene. The *CCND1* gene has subsequently been shown to be the same as the *BCL1* gene and its protein product is cyclin D1. *CCND1* is involved in the G_1 to S transition in the cell cycle [9 and references therein] (see Chapter 4).

A translocation between chromosomes 14 and 18 is found in approximately 85% of follicular lymphomas and involves the oncogene *BCL2* on chromosome 18, placing it next to the immunoglobulin heavy chain locus. Progression of follicular lymphomas to a more aggressive form is frequently associated with overgrowth of subclones containing chromosomal changes, in addition to the characteristic 14;18 translocation. In some cases this is associated with an 8;14 translocation involving the *MYC* gene, as seen in Burkitt's lymphoma. The *BCL2* translocation is not entirely specific for follicular lymphomas; it has also been reported in occasional cases of chronic lymphocytic lymphoma (CLL), ALL, and small noncleaved lymphoma, as well as in diffuse large-cell lymphomas and a variable number of Hodgkin's disease cases. The presence of the translocation may indicate a derivation from follicular lymphoma in these last two situations; in other neoplasms, however, it may represent a common first step to tumorigenesis, which may be followed by unique additional events [8 and references therein]. BCL2 protein expression similarly lacks diagnostic specificity because of its ubiquity. This protein is normally expressed in hematopoietic precursors, basal cells and in long-lived cells, such as medullary thymocytes and peripheral T cells and the B cells of the mantle zone of the follicles. The protein is not expressed in reactive germinal centers, a feature of diagnostic importance when compared with the strong immunostaining of neoplastic follicles. In addition to the follicular lymphomas, overexpression of BCL2 protein has been reported in diffuse lymphomas of both T- and B-cell lineages and both small- and large-cell type, CLL, ALL, plasma cell neoplasms, myeloid dysplasia and Hodgkin's disease. There must, therefore, be other mechanisms, besides the t(14;18) translocation, that can lead to BCL2 overexpression, either by deregulating the oncogene or by acting at the translational or post translational level. One of these mechanisms is the upregulation of BCL2 protein expression by EBV [10].

Genes homologous to *BCL2* play a key role in regulating apoptosis. *BCL2* and its relatives *BCLX* and *BAX* encode intracellular membrane-bound proteins that share homology in three domains with a wider family of viral and cellular proteins. The BCL2 and BCLX proteins enhance the survival of lymphocytes, and other cell types, but do not promote their proliferation. High levels of BAX or of a smaller BCLX variant antagonize the survival function of BCL2. The mechanism by which BCL2 promotes cell survival is unknown at present. However, the risk of B lymphoid tumors is enhanced probably because BCL2 can countermand the apoptotic action of other oncoproteins such as MYC. Expression of BCL2 may also constitute a major barrier to the success of genotoxic cancer therapy [11 and references therein].

Diffuse B-NHLs with t(14;18)(q32.3;q21.3) have a worse prognosis than those without this translocation and the development of a sensitive PCR assay now permits the detection of one NHL cell in 10^5 normal cells, making the screening of bone marrow for residual malignant cells after therapy feasible [12].

10.5 *ABL*

A 9;22 translocation is found in approximately 90% of all cases of CML. It is not diagnostic for this disease, as it is also present in 25% of adult ALLs, in 2–10% of childhood ALLs and in occasional cases of AML (<1%). The breakpoint on chromosome 9 occurs on the 5′ side of the first coding exon of the *ABL* proto-oncogene. On chromosome 22, the breakpoint occurs within the breakpoint cluster region (*BCR*) [or major cluster region (*MCR*), as it is now called]. This 5.8 kb DNA region contains four small exons and is part of the *PHL* gene. As a result of the translocation, a fusion gene is produced which, following splicing, is transcribed to give an 8.5 kb mRNA encoding a 210 kDa fusion protein. This altered protein exhibits elevated constitutive protein-tyrosine kinase activity (*Figure 10.2*). Molecular analysis of ALL reveals a second class of rearrangement in some cases. Here, the breakpoint in the *PHL* gene occurs near the 5′ end, resulting in a 7 kb chimeric mRNA encoding a 190 kDa protein which again has altered tyrosine kinase activity (*Figure 10.2*). The production of these fusion proteins is likely to be an important step in the pathogenesis of these diseases but it is still not clear whether the translocation is the primary event in tumorigenesis.

Although the Ph′ chromosome can be detected cytogenetically, molecular techniques are now being used to aid diagnosis, and many cases previously thought to be Ph′ negative have now been shown by Southern blotting to contain the rearrangement [13] (*Figure 10.3*). The accurate detection of this marker is important because patients who are Ph′ negative generally show reduced survival rates. Molecular techniques have also been used to detect the presence or absence of specific *BCR* exons. The presence or absence of exon 3 of the *BCR* gene accounts for some of the variability in the disease duration, thereby influencing the timing of onset of blast crisis. Patients with Ph′ chromosomes containing exon 3 were initially found to have a statistically shorter disease duration and more rapid onset of blast crisis [14]. However, more recent evidence indicates that the presence or absence of *BCR* exon 3 is not relevant to survival.

The ability to detect the presence of a *BCR–ABL* fusion gene by PCR has also allowed the development of a very sensitive diagnostic test for CML (*Figure 10.4*). In addition, it has been used successfully in the detection of the presence of residual disease following bone marrow transplantation. This might permit early therapy for patients with residual disease and thereby help to sustain remission. The major problem with this technique is also the reason

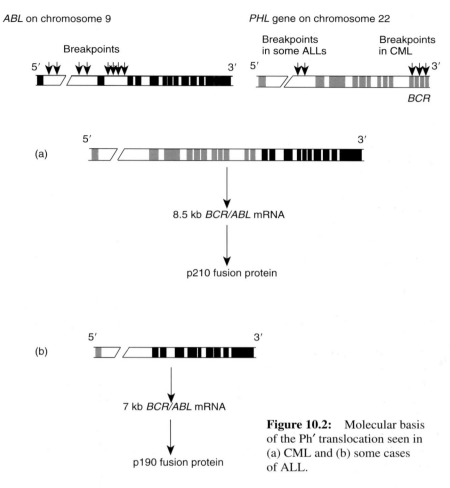

Figure 10.2: Molecular basis of the Ph′ translocation seen in (a) CML and (b) some cases of ALL.

for its use, that is, its high sensitivity. The persistence of low levels of Ph′-positive cells has been a consistent finding and it is not yet clear how many of these cells are necessary to re-establish disease. In the near future long-term follow-up of these patients will hopefully clarify this point and allow post-transplant therapy to be initiated where necessary.

10.6 *MDM2*

Recently, the murine double minute-2 (*MDM2*) gene was studied in 60 patients with B-cell chronic lymphocytic leukemia (B-CLL) or B-NHL. *MDM2* gene expression was low in normal B cells, whereas 28% of the patients with B-CLL or NHL had more than 10-fold higher levels of *MDM2* gene expression. This overexpression was found more frequently in patients with the low-grade type of lymphoma (56%) than in those with intermediate to high-grade types (11%)

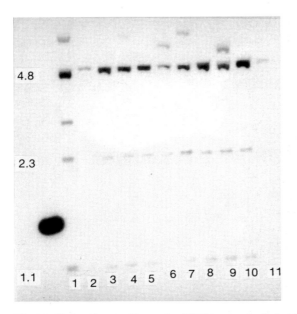

Figure 10.3:   Autoradiograph of DNA samples hybridized with the probe BCR-3 showing rearrangements in Ph' positive patients. Three normal bands are seen with molecular weights of 4.8 kb, 2.3 kb and 1.1 kb. If a translocation has occurred, one or two additional bands are present. Lanes 2, 3, 5, 8, 10 and 11 show no evidence of a rearrangement. Abnormal bands are seen in CML patients in lanes 1, 6, 7 and 9. An abnormal pattern is also seen in the patient with AML in lane 4.

and was found significantly more frequently in patients at advanced clinical stages. These results suggest that *MDM2* gene overexpression may play an important role in the tumorigenicity and/or disease progression of CLL and low-grade lymphomas of B cell origin [16].

10.7   *ALL1* and 11q23 translocations

A gene, variously designated as *ALL1*, *MLL*, *HRX* or *HTRX*, has been shown to have a critical role in multiple leukemic groups, as well as in occasional lymphomas, primarily by forming a fusion gene with many partners from different chromosomes resulting in the synthesis of chimeric RNAs and most likely of chimeric proteins. Alterations in the *ALL1* gene are found in 5–10% of all human leukemias and the *ALL1* gene is rearranged in the vast majority of 11q23 abnormalities. The various acute leukemias, both lymphoid and myeloid, with cytogenetic and molecular evidence for alterations in *ALL1*, uniformly carry a poor prognosis [17]. The t(4;11) translocation is the most common genetic alteration in early childhood leukemia and this, along with other rearrangements of the *ALL1* gene, sometimes submicroscopic, apparently contributes to at least 70% of the acute leukemias occurring in children

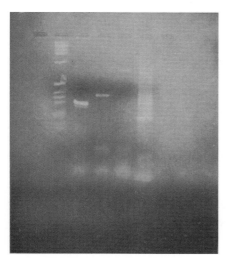

Figure 10.4: PCR analysis of two patients with CML plus a normal control. Lane 1: a 1 kb ladder size marker. Lane 2: a patient with a rearrangement present in whom *BCR* exon 2 is fused to *ABL* exon 2 (310 bp). Lane 3: rearrangement present with *BCR* exon 3 fused to *ABL* exon 2 (385 bp). Lane 4: normal control showing a single 808 bp band representing the normal *BCR* product. See ref. 15 for details of the primers. Photograph courtesy of Ben Disley and David Bourn, DNA Laboratory, Birmingham Heartlands Hospital, UK.

under the age of 1 year. Secondary leukemias wih 11q23 rearrangements typically have a myeloid phenotype and, as with the *de novo* leukemias, these are clinically aggressive and respond poorly to therapy.

10.8 Cyclin-dependent kinase 4 inhibitor, *CDKN2*

Approximately 10% of patients with ALL have cytogenetic abnormalities involving 9p21–22, the locus encoding the putative tumor suppressor gene *CDKN2* (*p16, MTS1*) (see Chapter 4). Homozygous deletion seems to be a major mechanism for activating *CDKN2* in ALL, CLL, malignant lymphoma, lymphoid transformation of chronic myelogenous leukemia and adult T-cell leukemia. Less frequently, mutation of one allele and concomitant loss of the remaining allele is responsible for *CDKN2* inactivation, and this occurs preferentially in T phenotype ALL [18 and references therein]. Whatever the mechanism, it appears that the disruption of 9p21 is an early event in the development of leukemia [19].

The *p15* (*MTS2*) gene is also a putative tumor suppressor located at chromosome 9p21 and it encodes a protein closely related to p16; p15 [20, 21]. Recent studies have shown that inactivation of either the *p15* or *p16* gene by point mutation is a very uncommon event in acute leukemias [21] and that the translocations affecting band 9p21 can participate in the inactivation of *p16* [22].

10.9   *p53*

p53 mutations occur moderately often in hematological malignancies and are particularly associated with progression of disease in both lymphoid and myeloid leukemias and lymphomas. In addition, *p53* mutations occur very frequently in Burkitt's and other high-grade B-cell lymphomas. Also, the Reed–Sternberg cells in Hodgkin's lymphoma express high levels of mutant *p53*, suggesting a major contribution of these cells to the disease [23]. A high incidence of *p53* gene mutation has also been found in mucosa-associated lymphoid tissue (MALT) lymphomas suggesting that *p53* plays an important role in the development of these lymphomas, particularly in high-grade transformations [24]. There is also a close association between *p53* abnormality and the replication error (RER$^+$) phenotype, suggesting that the two affect each other [24]. The prognostic significance of these associations has yet to be determined.

10.10   *RBTN* T-cell oncogenes

RBTN1 (*TTG1*) is located on chromosome 11p15 and is activated by the t(11;14)(p15;q11) translocation (see *Table 10.2*). *RBTN2* is located on 11p13 and is activated by the translocation t(11;14)(p13;q11). *RBTN3* is located on chromosome 12p12–13 and has not yet been found to be associated with a chromosomal translocation. The *RBTN* gene family encodes proteins with the cysteine-rich LIM domain that is similar to the electron transport protein ferredoxin and to zinc finger proteins. The LIM domain has been found in a number of proteins that are thought to be involved in developmental regulation [2]. While it is assumed that the *RBTN* family of genes are T-cell oncogenes that are activated by chromosomal translocations, the only direct evidence that *RBTN1* and *RBTN2* can participate in tumorigenesis has been obtained from studies in transgenic mice [2].

10.11   Detection of minimal residual disease

Detection of minimal residual disease (MRD) by cytogenetic analysis and the measurement of surface or immunological markers using flow cytometry has improved the sensitivity over histochemical techniques. However, the PCR has revolutionized the sensitivity of detection as chromosomal rearrangements can now be detected with a theoretical sensitivity of one malignant cell in a population of 10^5–10^6 cells [3 and references therein]. The Ig and/or TCR clonal rearrangements present at diagnosis in lymphoid leukemias and lymphomas provide a useful target which marks the abnormal cell progeny. Following PCR amplification of one or more of the clonal antigen receptor gene rearrangements present at diagnosis, a clone-specific probe can be produced to aid in the detection of residual cells in post-therapy samples.

However, the extreme sensitivity of the PCR brings its own problems including the risk of contamination resulting in false positive results. Questions must also be addressed concerning the nature of the residual cells harboring the disease-specific marker. Many patients are PCR positive after achieving clinical remission status but progress to become PCR negative. Are PCR positive cells present in a biopsy from a patient during clinical remission, clonogenic? It may not yet be ethical to intervene with additional therapy in patients who are judged PCR positive but who have no other clinical manifestations of disease [25]. Some CML patients, for example, can remain PCR positive for *BCR–ABL* without hematological relapse for years. Efforts are being directed at establishing more quantitative assays in order to determine not only whether residual or disseminated disease is present, but also at what level [25].

The potential value of PCR analysis can be illustrated by the following examples. In CML and ALL there is evidence that transient *BCR–ABL* positivity is not usually followed by hematological relapse, while patients who have serial samples which are positive have a high risk of relapse [26]. In Philadelphia-positive ALL an increase in *BCR–ABL* transcript numbers on sequential analysis after transplantation permits the early identification of individual patients who are likely to progress to cytogenetic and hematological relapse [27].

However, while molecular analysis of remission specimens can predict a significant fraction of clinical relapses, the clinical value of MRD detection at this level of sensitivity still remains unclear. The hope is that the results of molecular analysis will enable the implementation of new treatment strategies for patients whose remission marrows remain positive. To clarify the role of molecular analysis in the detection of MRD, prospective studies are needed [3].

References

1. Parker, S.L., Tong, T., Bolden, S.B. and Wingo, P.A. (1996) *CA – Cancer J. Clinic.*, **46**, 5.
2. Rabbitts, T.H. (1993) *Biochem. Soc. Trans.*, **21**, 809.
3. Boxer, L.M. (1994) *Annu. Rev. Med.*, **45**, 1994,
4. Look, A.T. (1995) *Adv. Cancer Res.*, **67**, 25.
5. Klein, G. and Klein, E. (1986) *Cancer Res.*, **46**, 3211.
6. Sweetenham, J.W. (1994) *Exp. Hematol.*, **22**, 5.
7. Bos, J.L. (1989) *Cancer Res.*, **49**, 4682.
8. Lim, L-C., Segal, G.H. and Wittwer, C.T. (1995) *Am. J. Clin. Pathol.*, **104**, 689.
9. Schuuring, E. (1995) *Gene*, **159**, 83.
10. Inghirami, G. and Frizzera, G. (1994) *Am. J. Clin. Pathol.*, **101**, 681.
11. Cory, S. (1995) *Ann. Rev. Immunol.*, **13**, 513.
12. Dyer, M.J.S. (1995) in *Oncology. A Multidisciplinary Textbook* (A. Horwich, Ed.). Chapman & Hall Medical, London, p. 585.
13. Blennerhassett, G.T., Furth, M.E., Anderson, A. *et al.* (1988) *Leukemia*, **2**, 648.
14. Grossman, A., Scheer, R.T., Arlin, Z. *et al.* (1989) *Am. J. Hum. Genet.*, **45**, 729.
15. Cross, N.C.P., Melo, J.V., Feng, L. and Goldman, J.M. (1994) *Leukemia*, **8**, 186.

16. Watanabe, T., Hotta, T., Ichikawa, A., Kinoshita, T., Nagai, H., Uchida, T., Murate, T. and Saito, H. (1994) *Blood*, **84**, 3158.
17. Canaani, E., Nowell, P.C. and Croce, C.M. (1995) *Adv. Cancer Res.*, **66**, 213.
18. Nakao, M., Yokota, S., Kaneko, H., Seriu, T., Koizumi, S., Takaue, Y., Fujimoto, T. and Misawa, S. (1996) *Leukemia*, **10**, 249.
19. Ragione, F.D., Mercurio, C. and Iolascon, A. (1995) *Haematologica*, **80**, 557.
20. Pines, J. (1995) *Adv. Cancer Res.*, **66**, 181.
21. Sill, H., Aguiar, R.C.T., Schmidt, H., Hochhaus, A., Goldman, J.M. and Cross, N.C.P. (1996) *Br. J. Haematol.*, **92**, 681.
22. Duro, D., Bernard, O., Valle, V. D., Leblanc, T., Berger, R. and Larsen, C.-J. (1996) *Cancer Res.*, **56**, 848.
23. Immamura, J., Miyoshi, I. and Koeffler, H.P. (1994) *Blood*, **84**, 2412.
24. Peng, H., Chen, G., Du, M., Singh, N., Isacson, P.G. and Pan, L. (1996) *Am. J. Pathol.*, **148**, 643.
25. McCarthy, K.P. and Wiedman, L.M. (1995) in *PCR Applications in Pathology. Principles and Practice* (D.S. Latchman, Ed.). Oxford University Press, Oxford. p. 216.
26. Gaiger, A., Lion, T., Kahls, P. *et al.* (1993) *Leukemia*, **1**, 1766.
27. Van Rhee, F., Marks, D.I., Lin, F. *et al.* (1995) *Leukemia*, **9**, 329.

Chapter 11

Other cancers

11.1 Introduction

Many of the genes discussed in Chapters 2–4 have a clear diagnostic use but have not been considered in the chapters on cancers according to the site of origin. In many cases the tumors these genes cause are rare. However the ability to detect mutations in the genes, to be able to offer presymptomatic and, in some cases, prenatal diagnosis, has led to the development of screening protocols so that unnecessary screening can be avoided for those at low risk of developing the cancer. For those at high risk, careful monitoring should allow early treatment and hopefully prevent early death from cancer. These rarer cancers can therefore provide models for use in other cancers in the future as the role of genes in the development of common cancers becomes known.

11.2 Retinoblastoma

The ability to accurately predict who is at high risk in retinoblastoma families is important. Those identified to be at high risk can be screened whereas low-risk individuals can be removed from a very intensive screening program which involves regular and intensive retinal examination. In familial retinoblastoma, accounting for around 10–15% of cases, it is possible to use linked polymorphic markers to track the high-risk chromosome through families. In the remaining 85–90% of sporadic unilateral cases, approximately 15% will carry a germ-line mutation. Identifying the causative mutation in them is the only way, other than by clinical examination, in which their sibs and offspring can be tested to determine their risks.

Approximately 5% of cases of retinoblastoma have a cytogenetically visible deletion and around 10% of cases can be detected by Southern blotting. In the remainder, the mutations are small insertions, deletions or point mutations [1,2] distributed throughout the gene (*Figure 11.1*) [3]. Over 70% of *RB1* mutations result in a truncated protein product. C to T transitions at CGA codons, of which there are 14 in the *RB1* gene, have been demonstrated in around 15–18% of cases [4]. Complete sequencing of the coding region of the gene has been carried out but this is an expensive and time-consuming procedure and it is not

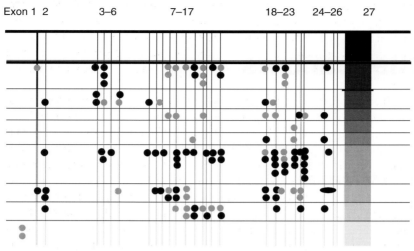

Figure 11.1: Distribution of mutations throughout the *RB1* gene. ● Mutations detectable by fragment analysis of size or copy number of exons. ● Single nucleotide substitutions requiring other methods of detection. Figure reproduced from ref. 3.

clear if all mutations occur within the screened region (see below). In retinoblastoma, mRNA from the mutant allele is not present in lymphocytes at a high enough level to allow cDNA to be used for analysis, so step-by-step analysis of each exon is therefore the only possibility. Several prescreening techniques such as SSCP, DGGE and RNase protection have been used to detect *RB1* mutations. All show a disappointingly low pick-up rate of 20–50% and are still labor intensive because of the large size of the gene. The low pick-up rate may in part be due to the presence of mutations in introns, promoter regions or other upstream regulatory sequences. In addition, other mutational mechanisms may exist. For example, hypermethylation has been suggested to occur in *RB1* which may inactivate the gene [5]. Methods to improve mutation detection in *RB1* are therefore required in order to alter the risks in all or even the majority of patients.

Some correlation has been made between genotype and phenotype in retinoblastoma. A mild phenotype has been associated with missense mutations [1] and has also been seen in cases of in-frame deletions [6]

11.3   Wilms' tumor

Mutations in the *WT1* gene are found in some Wilms' tumors and in the germline of patients with an inherited susceptibility. Mutations in tumors, including large and small deletions, and point mutations have been described [7]. Mutations in the zinc finger domain result in loss of DNA binding activity whereas those in the transactivation domain convert the protein into a transcriptional activator rather than a repressor. Ten percent of Wilms' tumors have a cytogenetically detectable deletion involving one chromosome. Approximately 10% of Wilms' tumors express an aberrantly spliced transcript associated with an in-frame deletion of exon 2, the result of which is to convert the protein from a transcriptional repressor to a transactivator of its own DNA target [8].

Germ-line mutations in *WT1* are rare. Cytogenetically visible deletions at 11p13 have been reported in WAGR syndrome as described in Section 3.3.1, a finding which led to the initial localization of the gene. Missense and nonsense mutations have been described in some patients [7]. Females with germ-line mutations generally do not exhibit salient genitourinary anomalies, suggesting that *WT1* is more important for the development of the male genital tract than the female. Some germ-line mutations cause the severe condition known as Denys–Drash syndrome (DDS; see Section 3.3.1). Mutations in this condition are found near or in the zinc finger coding region of the gene. One mutation, a C to T transition at codon 394 of exon 9 has been described in approximately half of the cases [9]. Other mutations are frequently found in exon 9 suggesting that this is a hot-spot for mutations in DDS. There is marked phenotypic variation seen between DDS patients with the same common mutation and it has also been described in a female with isolated Wilms' tumor and no other genitourinary abnormalities.

Wilms' tumors are also seen in Beckwith–Wiedemann syndrome (BWS) as discussed in Section 3.3.1, a condition associated with chromosome 11p15. Approximately 2–3% of BWS patients have a chromosomal abnormality, either a paternal duplication of 11p15 or a maternally derived inversion or balanced translocation. About 20% of sporadic cases of BWS show uniparental disomy (UPD). These results have implicated an imprinted gene in the pathogenesis of BWS and a candidate gene is *IGF2*. The detection of these abnormalities can be helpful in the confirmation of the disease particularly in cases with less apparent clinical features and can then allow accurate counseling of patients. UPD has been associated with an increased risk of Wilms' tumor in some studies and if confirmed may help target screening to a high-risk group. The pattern of allele loss in Wilms' tumors suggests that another more conventional tumor suppressor gene, *WT2*, may also be located at 11p15.

Finally linkage analysis in some WT families has identified the existence of a *WT3* gene as chromosome 11 has been excluded in these families [10]. However to date this locus has not been mapped.

11.4 Neurofibromatosis type 1 (NF1)

Many polymorphic markers are known in the *NF1* gene meaning that linkage analysis is possible in some families [11]. However, as *NF1* has one of the highest new mutation rates of any gene, direct detection of mutations, usually by SSCP, DGGE or heteroduplex analysis, is required for the remainder.

Detection of mutations in *NF1* has proved as difficult as for *RB1* [12]. Again the gene is large, with a wide distribution of mutations throughout the gene, most of which are unique to one family. Only one 'common' mutation has been described which is a C to T transition in exon 31 but even this is relatively rare having only been reported six times. The nature of mutations is similar to other tumor suppressor genes including single base substitutions, deletions of all sizes, insertions and three reported cytogenetic rearrangements. Over 100 different mutations have so far been described [12] with approximately 82% causing truncation of the protein. This, together with the availability of mRNA for this condition, has meant that PTT across several exons has recently been used to detect mutations [13]. Several cases have been described in which a reduction in the expression of one allele was observed [14]. To date, because of the low number of mutations identified and the phenotypic variation within families themselves, it has not been possible to show any correlation between genotype and phenotype. It is not yet clear whether the three genes, *EVI2A*, *EVI2B* and *OMPG*, embedded in intron 27 of the *NF1* gene could play a role in the variable phenotype [12].

At present, because of the low number of identifiable mutations, molecular analysis has only been available to a small number of families. Presymptomatic diagnosis has not been requested often and it is relatively easy to make a clinical diagnosis even in early childhood. As it is not possible to predict disease severity or prognosis in this condition, and the phenotype can vary so widely, prenatal diagnosis has rarely been requested. If it is possible to make such correlations in the future, this service may be in higher demand from families.

11.5   Neurofibromatosis type 2 (NF2)

In suitable families, linkage analysis for both prenatal and presymptomatic diagnosis of NF2 is possible and many polymorphic markers have been described [15]. However, as around 50% of cases are new mutations, mutation analysis is important in this disorder. Mutation analysis in NF2 has been much more successful than in NF1, probably because of the smaller size of the gene and both SSCP and DGGE have been used [16,17]. Mutations are usually unique to families as with the other tumor suppressor genes although C to T transitions particularly at CGA codons such as codon 57 have been reported on more than one occasion [18]. Mutations include deletions, insertions, point mutations and interference with normal splicing, the majority of mutations being truncating (*Figure 11.2*) [16]. Missense mutations are relatively rare in the *NF2* gene. The ability to detect mutations in *NF2* means that presymptomatic screening of at-risk members of families can be used to improve diagnosis and reduce regular radiographic and audiological screening. No absolute correlation can be made between genotype and phenotype as clinical manifestations vary between patients with the same mutation although the disease is usually of a similar severity within families. However a milder phenotype has been associated with mutations which preserve the integrity of the carboxyl-terminal portion of the protein [17] and has also been noted in patients with a complete gene deletion [15]. A mild phenotype has also been noted in one patient who was a somatic mosaic for an *NF2* mutation [19].

Schwannomas develop as part of the NF2 phenotype and also occur sporadically. Inactivating mutations have been found in the *NF2* gene, both in the sporadic tumors as well as in those from patients with NF2 [20]. In contrast to the germ-line mutations, the majority of somatic mutations are deletions [16]. Although there is no hot-spot for mutations, approximately 73% have been found in the first eight exons [20]. Inactivation of the second copy of the gene has been found in the tumors, indicating that both copies of the gene need to be lost for tumorigenesis. These results confirm the role of *NF2* in the development of schwannomas although the possibility that another gene is also involved in sporadic schwannomas cannot be excluded as mutations in *NF2* have not been found in around 50% of cases. Somatic mutations in *NF2* have also been found in sporadic meningiomas together with LOH of the second copy implicating the gene in the development of this tumor as well [21].

SOMATIC MUTATIONS

	1	2	3	4	5	6	7	8	9	10	11	12	13	14	15	16	17
Deletion																	
Frameshift	1	8		2	1		3	6	1	5	3	6		2	3		
Splice			2				1	1		1	1				1		
In-frame																	
Insertion																	
Frameshift							1	1					1				
Splice							1										
Point mutant																	
Missense		1					1										
Nonsense		3					4	3			2	3					
Splice		1		1			3	2									

Exon no.	1	2	3	4	5	6	7	8	9	10	11	12	13	14	15	16	17

	1	2	3	4	5	6	7	8	9	10	11	12	13	14	15	16	17
Deletion																	
Frameshift		1	1					1		2	2	1		2			
Splice																	
In-frame			1														
Insertion																	
Frameshift								1	1								
Splice											2						
Point mutant																	
Missense								1			1	1					
Nonsense		5				1		3		1	4	1	3		2		
Splice		3			3	1				1		2			1		

GERM-LINE MUTATIONS

Figure 11.2: Summary of mutations in the *NF2* gene. The gene is shown in the center of the figure with the exons numbered. Somatic changes are summarized above the gene and germ-line mutations below. The numbers correspond to the number of times each mutation has been reported. Figure reproduced from ref. 16 with permission from The University of Chicago Press.

11.6   Multiple endocrine neoplasia

Multiple endocrine neoplasia type 1 (MEN1) is associated with the development of parathyroid, pancreatic and anterior pituitary tumors. Multiple endocrine neoplasia types 2A and 2B (MEN2A and MEN2B) and familial medullary thyroid carcinoma (FMTC) are three clinically distinct syndromes inherited in an autosomal dominant manner but which share the common feature of medullary thyroid cancer. In addition, MEN2A and MEN2B are associated with phaeochromocytomas. The causative mechanisms for these conditions have recently been elucidated, making screening in families a realistic possibility.

11.6.1   MEN1

The gene for MEN1 was mapped to chromosome 11 following LOH studies of markers spread throughout the genome. Subsequently linkage analysis in families localized the gene to 11q13. This meant that presymptomatic diagnosis for 'at-risk' members of families became a possibility for this condition particularly using polymorphic markers at the muscle glycogen phosphorylase locus [22] and thereby enabling modification of the extensive screening protocols.

So far the gene for MEN1 has not been cloned. However LOH in pancreatic and parathyroid tumors suggests that it will be a tumor suppressor gene. Sporadic parathyroid, pancreatic and pituitary tumors have also been shown to have deletions of chromosome 11, suggesting that the *MEN1* gene may be involved in their development.

11.6.2   MEN2

The gene for MEN2A was assigned to chromosome 10 by linkage analysis in 1987 [23]. Subsequently both MEN2B and FMTC were also mapped to the same region [24]. Extensive family studies narrowed the region in which the gene lay, to the pericentromeric region of chromosome 10 and, as with other conditions discussed in this chapter, immediately meant that families could be offered presymptomatic diagnosis using polymorphic linked markers.

It was initially assumed that the *MEN2* gene would be a tumor suppressor so LOH studies were initiated to help further map the region containing the gene. However no evidence of any allele loss was found on chromosome 10 suggesting that one possibility for the cause of MEN2 could be a dominantly acting gene rather than a recessive tumor suppressor. A candidate gene, which had also been mapped to the pericentromeric region of chromosome 10, was the *RET* proto-oncogene. The product of this gene is a receptor tyrosine kinase, activation of which results in transduction of signals for cell proliferation. This gene was analyzed for the presence of mutations in individuals with MEN2A and FMTC using chemical mismatch cleavage and SSCP and mutations were identified in the majority of cases. In MEN2A and FMTC, mutations are found primarily in the cysteine-rich extracellular domain and most are substitutions of cysteine with another amino acid [24]. Mutations involving codon 634 occur in up to 75% of MEN2A/FMTC families. Some genotype–phenotype correlations have been suggested. The most common mutation in MEN2A, associated with parathyroid cancer, is the substitution of cysteine at codon 634 with arginine [25] suggesting that this mutation may particularly predispose to this cancer. However, identical mutations have been shown to be causal in both MEN2A and FMTC suggesting that other modifiers of the phenotype must exist. In MEN2B a different mutation, a methionine to threonine mutation at codon 918, is causative in the majority of cases. This mutation occurs in the tyrosine kinase domain of *RET* [26]. Sporadic medullary thyroid cancers have also been shown to have mutations in *RET*.

The ability to detect mutations in these three groups of disorders means that molecular testing can be used to identify high-risk patients and those at low risk can be removed from unpleasant as well as costly screening protocols. Patients with mutations are recommended to undergo total thyroidectomy to prevent the occurence of medullary thyroid cancer and to continue to be screened for phaeochromocytomas.

11.7   Familial melanoma

Cutaneous malignant melanoma is a potentially fatal form of skin cancer whose incidence is rising in many regions of the world. In the UK, the incidence of the disease is 4 in 100 000 males and 7 in 100 000 females, and the incidence has been rising at a rate of over 7% per annum. In the US, the incidence of this cancer is rising faster than any other, and in Australia the lifetime risk of melanoma is 5%. The ability to detect causative genes in this disorder is likely to help us to understand how the tumor develops. A number of nonrandom karyotypic changes have been identified in the progression from the normal melanocyte to a malignant melanoma and have been shown to involve primarily chromosomes 1, 6, 7, 9 and 10 [27]. Approximately 8–10% of cases of melanoma occur in patients with a familial predisposition, often seen in association with dysplastic naevi, a lesion considered to be a precursor of melanoma.

The first candidate region for a familial melanoma gene was chromosome 9p, identified because of consistent LOH in this area and because of a case of a cytogenetic abnormality involving chromosomes 9p and 5p in an individual with multiple melanoma. Subsequently a melanoma predisposing gene was assigned to 9p by linkage studies in 11 families [28]. A gene within the candidate region was identified as the p16 or cyclin-dependent kinase inhibitor 2 (*CDKN2*) gene [29]. This gene encodes a protein whose product inhibits the cyclin-dependent kinases, CDK4 and 6, and subsequently inhibits their ability to phosphorylate the retinoblastoma protein (see Section 4.4.1). Any inactivation of p16 would therefore lead to loss of the retinoblastoma control of the G_1

checkpoint and continuous cell proliferation (see *Figure 4.3*). Germ-line mutations, both truncating and missense, have been reported in *CDKN2* in a proportion of melanoma families confirming that this is a causative gene [30].
However familial melanoma has been shown to be a heterogenous disorder,
with at least one other gene involved mapping to chromosome 1p. This gene
has not yet been identified [31]. A recent study has also identified a mutation in
the *CDK4* gene itself in two unrelated families [32]. This mutation occurs in
the p16 binding domain of *CDK4*, thereby generating a gene which is resistant
to the inhibitory effects of p16, which acts as a dominant oncogene.

11.8   Tuberous sclerosis

Tuberous sclerosis is an autosomal dominantly inherited condition with multi-
system abnormalities characterized by hamartomas in different organs including skin, kidney, brain, heart, retina and lung. In around two-thirds of cases the
disease appears to be sporadic. The disorder is genetically heterogeneous.
Approximately 30–50% of families have been shown to be linked to 9q34 [33]
with the majority of the remainder being linked to a locus on chromosome 16
[34]. LOH has been demonstrated at both loci, indicating that the causative
genes were likely to be tumor suppressor genes [35].

The gene on chromosome 16, *TSC2*, has been cloned and shown to be
widely expressed [36]. Its product, called tuberin, has a region homologous to
the GTPase activating protein GAP3. The gene on chromosome 9, *TSC1*, has
still to be isolated. SSCP has been used to look for mutations in the *TSC2* gene
[37]. These included missense and nonsense mutations, small deletions and a
tandem duplication, scattered throughout the gene. In addition, large deletions
and intragenic deletions have been described [36]. No correlation has yet been
found between the nature of mutation and clinical severity but the number of
mutations identified so far is small. LOH was found more frequently at the
TSC2 locus in hamartomas from patients with sporadic disease, compared to
the *TSC1* locus. As this group had the more severe organ impairment, it was
suggested that the *TSC2* locus might confer a greater risk of early kidney failure or more rapid growth of giant cell astrocytomas.

11.9   Childhood cancers

11.9.1   *Neuroblastoma*

Neuroblastoma is the most common solid tumor in childhood. The mean age at
diagnosis in sporadic cases is 30 months and drops to 9 months for familial
cases. Cytogenetic and molecular studies have implicated inactivation of a
tumor suppressor gene on chromosome 1 and activation of an oncogene as
being involved in this disease. Cytogenetic investigations showed that DMs
and HSRs were found in association with neuroblastomas. These regions were

shown to contain a DNA sequence with partial homology to the *NMYC* gene, and both neuroblastoma cell lines and tumors have been shown to carry amplified *NMYC*. Subsequently similar results were found in other tumors such as SCLC and retinoblastoma. Experimental systems have been used to demonstrate that high expression of *NMYC* can modulate cell growth.

NMYC was one of the first oncogenes demonstrated to be of clinical significance. Stage III and IV neuroblastomas have a poor prognosis with only 10–30% survival at 2 years. A strong correlation has been found between stage III and IV disease and *NMYC* amplification. In contrast, stage IV tumors which frequently regress have rarely been associated with *NMYC* amplification [38]. *In situ* hybridization has been used to detect increased levels of *NMYC* mRNA and correlates well with determination of gene copy number. This technique can be used to detect increased expression of *NMYC* in a largely negative cell population where direct analysis of amplification would give a negative result.

LOH has been identified in neuroblastomas and localized to 1p36, implicating this region as the site of a tumor suppressor gene involved in development of this tumor. Further studies have suggested that two suppressor genes are located in this region. A more proximal suppressor is associated with *NMYC* amplification while a distal, possibly imprinted, gene is involved in cases without amplification [39].

11.9.2   *Rhabdomyosarcoma*

Soft tissue sarcomas account for 4–8% of childhood cancers. Rhabdomyosarcoma is the most common soft tissue sarcoma in persons below the age of 21 years. Embryonal rhabdomyosarcomas account for approximately 60% of cases, usually occuring in younger children and are found located in the head and neck, orbit and genitourinary tract. Alveolar rhabdomyosarcomas often occur in adolescence in the extremities. Chromosomal abnormalities are associated with both types of tumor. In embryonal rhabdomyosarcomas, complex and variable changes are found. However in alveolar rhabdomyosarcomas a consistent translocation, t(2;13)(q35;q14) is found [40]. The breakpoint on chromosome 2 was shown to disrupt the developmental control gene, *PAX3*, whereas that on chromosome 13 was shown to be a member of the forkhead family of transcription factors (*FKHR*). The rearrangement creates a fusion gene containing the regions encoding the *PAX3* paired box and homeodomains together with the carboxyl-terminal region of the *FKHR* DNA binding domain and the unique *FKHR* carboxyl-terminal domain (*Figure 11.3*). The likely result of this is to produce a fusion protein which has the ability to bind normal *PAX3* target genes but with aberrant regulation of transcription through the *FKHR* regulatory sequences. Alternatively the translocation could increase the oncogenic potential of either of the genes by increasing their levels of expression.

Although no consistent chromosomal rearrangements are seen in embryonal

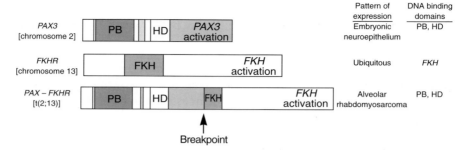

Figure 11.3: Schematic diagram of the results of the t(2;13)(q35;q14) translocation in alveolar rhabdomyosarcoma which results in the production of a fusion gene containing the paired box (PB) and homeodomain (HD) of *PAX3* and the bisected *FKHR* DNA binding domain and the *FKHR* activation domain. Figure reproduced from ref. 40.

rhabdomyosarcomas, studies have shown LOH of 11p15.5, implicating the presence of a tumor suppressor gene at this locus, which may be involved in tumor development [40]. Interestingly this region is implicated in Wilms' tumor development as discussed above and both rhabdomyosarcoma and Wilms' tumor are a feature of BWS.

11.9.3 Ewing's sarcoma

This cancer has a peak age of onset in adolescence. A t(11;22)(q24;q12) translocation has been associated with over 80% of cases. This again produces a fusion gene involving the *EWS* gene and the *FLI1* gene [41]. A second translocation, t(21;22) producing a second fusion gene involving *EWS* and *ERG* on chromosome 21 has also been found in a proportion of cases. Both *FLI1* and *ERG* are members of the ETS group of transcription factors. The *EWS* gene is also involved in malignant melanomas.

11.10 Ataxia telangiectasia

Ataxia telangiectasia (AT) is an autosomal recessive condition characterized by the development of cerebellar ataxia in the first decade. Cytogenetically it is characterized by an increased incidence of chromosome breaks and therefore belongs to the group of chromosome breakage syndromes.There is a 10–20% risk of developing a cancer, usually lymphoma or leukemia. In addition AT heterozygotes have been shown to be at increased risk of breast cancer and could account for up to 7% of cases [42]. The gene for this disorder has recently been isolated and shown to encode a protein with similarity to yeast and mammalian phosphatidylinositol-3′-kinases, which are involved in signal transduction and cell cycle control [43]. The isolation of this gene should mean that prenatal diagnosis and carrier detection can be offered to AT families. In addition, it will hopefully allow the role of this gene in the development of breast cancer to be elucidated.

References

1. Liu, Z., Song, Y., Bia, B. and Cowell, J.K. (1995) *Genes, Chrom. Cancer*, **14**, 277.
2. Blanquet, V., Turleau, C., Gross-Morand, M.S., Senamaud-Beaufort, C., Doz, F. and Besmond, C. (1995) *Hum. Mol. Genet.*, **4**, 383.
3. Gallie, B.L., Hei, Y.-J., Mostachfi, H. and Dunn, J.M. (1995) in *Molecular Genetics of Cancer* (J.K. Cowell, Ed.). BIOS Scientific Publishers, Oxford, p. 1.
4. Cowell, J.K., Smith, T. and Bia, B. (1994) *Eur. J. Cancer*, **2**, 281.
5. Sakai, T., Toguchida, J., Ohtani, N., Yandell, D.W., Rapaport, J.M. and Dryja, T.P. (1991) *Am. J. Hum. Genet.*, **48**, 880.
6. Dryja, T.P., Rapaport, J., McGee, T.L., Nork, T.M. and Schwarz, T.L. (1993) *Am. J. Hum. Genet.*, **52**, 1122.
7. Huff, V. and Saunders, G.F. (1993) *Biochim. Biophys. Acta*, **1155**, 295.
8. Haber, D., Park, S., Maheswaran, S. *et al.* (1993) *Science*, **262**, 2057.
9. Mueller, R.F. (1994) *J. Med. Genet.*, **31**, 471.
10. Grundy, P., Koufos, A., Morgan, K., Li, F.P., Meadows, A.T. and Cavanee, W.K. (1988) *Nature*, **336**, 374.
11. Jorde, L.B., Watkins, W.S., Viskochil, D., O'Connell, P. and Ward, K. (1993) *Am. J. Hum. Genet.* **53**, 1038.
12. Shen, M.H., Harper, P.S. and Upadhyaya, M. (1996) *J. Med. Genet.*, **33**, 2.
13. Heim, R.A., Kam-Morgan, L.N.W., Binnie, C.G. *et al.* (1995) *Hum. Mol. Genet.*, **4**, 975.
14. Hoffmeyer, S., Assum, G., Kaufman, D. and Krone, W. (1994) *Nature Genetics*, **6**, 331.
15. Watson, C.J., Gaunt, L., Evans, G., Patel, K., Harris, R. and Strachan, T. (1993) *Hum. Mol. Genet.*, **2**, 701.
16. MacCollin, M., Ramash, V., Jacoby, L.B. *et al.* (1994) *Am. J. Hum. Genet.*, **55**, 314.
17. Merel, P., Hoang-Xuan, K., Sanson, M. *et al.* (1995) *Genes, Chrom. Cancer*, **12**, 117.
18. Sainz, J., Figueroa, K., Baser, M.E., Mautner, V.F. and Pulst, S.-M. (1995) *Hum. Mol. Genet.*, **4**, 137.
19. Bourn, D., Carter, S.A., Evans D.G.R., Goodship, J. and Coakham, H. (1994) *Am. J. Hum. Genet.*, **55**, 69.
20. Bijlsma, E.K., Merel, P., Bosch, D.A., Westerveld, A., Delattre, O., Thomas, G., and Hulsebos, T.J.M. (1994) *Genes, Chrom. Cancer*, **11**, 7.
21. Papi, L., Rosaria De Vitis, L., Vitelli, F., Ammannati, F., Mennonna, P., Montali, E. and Bigozzi, U. (1995) *Hum. Genet.*, **95**, 347.
22. Thakker, R.V., Wooding, C., Pang, J.T. *et al.* (1993) *Ann. Hum. Genet.*, **57**, 17.
23. Mathew, C.G.P., Chin, K.S., Easton, D.F. *et al.* (1987) *Nature*, **328**, 527.
24. Goodfellow, P. (1994) *Curr. Opin. Genet. Dev.*, **4**, 446.
25. Mulligan, L.M., Eng, C., Healey, C.S. *et al.* (1994) *Nature Genetics*, **6**, 70.
26. Eng, C., Smith, D.P., Mulligan, L.M. *et al.* (1994) *Hum. Mol. Genet.,* **3**, 237.
27. Fountain, J.W., Bale, S.J., Housman, D.E. and Dracopoli, N.C. (1990) *Cancer Surv.*, **9**, 645.

28. Cannon-Albright, L., Goldgar, D.E., Meyer, L.J. *et al.* (1992) *Science*, **258**, 1148.
29. Kamb, A., Gruis, N., Weaver-Feldhaus, J. *et al.* (1994) *Science*, **264**, 436.
30. Walker, G.J., Hussussian, C.J., Flores, J.F. *et al.* (1995) *Hum. Mol. Genet.*, **4**, 1845.
31. Goldstein, A.M., Dracopoli, N.C., Engelstein, M., Fraser, M.C., Clark, W.H. and Tucker, M.A. (1994) *Am. J. Hum. Genet.*, **54**, 489.
32. Zuo, L., Weger, J., Yang, Q., *et al.* (1996) *Nature Genetics*, **12**, 97.
33. Haines, J.L., Short, M.P., Kwiatkowski, D.J. *et al.* (1991) *Am. J. Hum. Genet.*, **49**, 764.
34. Povey, S., Burley, M.W., Attwood, J. *et al.* (1994) *Ann. Hum. Genet.*, **58**, 107.
35. Carbonara, C., Longa, L., Grosso, E. *et al.* (1996) *Genes, Chrom. Cancer*, **15**, 18.
36. European Chromosome 16 Tuberose Sclerosis Consortium (1993) *Cell*, **75**, 1305.
37. Wilson, P.J., Ramesh, V., Kristiansen, A. *et al.*, (1996) *Hum. Mol. Genet.*, **5**, 249.
38. Broder, G.M. and Fong, C.T. (1989) *Cancer Genet. Cytogenet.*, **41**, 153.
39. Caron, H., Peter, M., van Sluis, P., Speleman, F., de Kraker, J., Laureys, G., (1995) *Hum. Mol. Genet.*, **4**, 535.
40. Shapiro, D. (1995) in *Molecular Genetics of Cancer* (J.K. Cowell, Ed.). BIOS Scientific Publishers, Oxford, p. 205.
41. May, W.A., Gishizky, M.L., Lessnick, S.L. *et al.* (1993) *Proc. Natl Acad. Sci. USA*, **90**, 5752.
42. Easton, D. Ford, D. and Peto, J. (1993) *Cancer Surv.*, **18**, 1.
43. Savitsky, K., Bar-Shira, A., Gilad, S. *et al.* (1995) *Science*, **268**, 1749.

Chapter 12

Therapeutic applications of oncogenes and their products

12.1 Introduction

Current nonsurgical approaches to the treatment of cancer suffer from the major limitation that they are not specific for malignant cells. While chemotherapy significantly improves survival rates in certain tumors, such as ALL and testicular cancer, for most 'solid tumors', which comprise the majority of the cancer incidence and mortality statistics (lung, breast, gastric, colon, ovary), survival figures have hardly improved over 30 years. In earlier chapters we presented compelling evidence for the importance of oncogenes in the etiology of many human tumors (see also ref. 1 for review). These genes and their encoded proteins are therefore potential targets for attack by specific therapeutic agents. In this chapter we will discuss whether this expectation has been, or is near to being, realized.

Potential targets for therapy include the oncogenes themselves, their RNA transcripts, and their protein products. Initially the expectation was that therapy directed against oncogenes or their products would be more specific to cancer and less harmful to normal tissues of the body. Since, in most cases, oncogene products have proved to be ubiquitous, specific inactivation of oncogenes would have wide-ranging effects. An example of this is the *SRC*-related protein kinases. These were among the first oncogene products to be recognized and specific inhibitors of these kinases seemed logical candidates for anticancer drugs. It is now known that the kinases carry out numerous functions essential to cell survival. Generalized interference with tyrosine phosphorylation would, therefore, have as low a therapeutic index as traditional anticancer drugs. In addition it is the lack of tyrosine substrate at the end of the cytoplasmic tails of kinase receptors that is responsible for their failure to autoregulate. The tyrosine kinase portions of these oncogene products seem to differ only slightly from their normal proto-oncogene counterparts. Thus, truly specific anti-ONC protein inhibitors may be difficult to realize in practice [2].

Several approaches to the therapeutic uses of oncogenes are worth considering:

(1) antibodies can be used against growth factors and other factors associ-

ated with the transformed phenotype, such as enzymes or proto-oncogene products like RAS p21. This includes the use of antibodies to focus toxic agents or cells of the immune system on cancer cells;

(2)   oncogenic nucleic acid sequences could be targeted by antisense oligonucleotides of DNA or RNA, or by nucleotide antimetabolites such as dideoxy- or methylphosphonate-modified nucleotides;

(3)   a mutated activated oncogene might be directly replaced by its nonmutated normal counterpart by gene therapy.

12.2   Role of antibodies

12.2.1   Use of antibodies as carriers of cytotoxic agents

The idea of targeting drugs to cancer cells using antibodies as carriers is not a new concept; it was proposed early in this century by Paul Ehrlich. However, it is only in the last 25 years that this approach has begun to be tested experimentally and a major impetus to its evaluation was the development of monoclonal antibodies by Kohler and Milstein in the mid 1970s [3]. As they are derived from the clonal expansion of a single antibody-producing cell, monoclonal antibodies are homogeneous and react with only a single antigenic determinant. The production of such antibodies was clearly a major advance over the polyclonal antisera available previously which contained mixtures of antibodies of varying specificities. In the clinical as well as the experimental setting, this lack of definition made polyclonal antisera less useful for purposes where high specificity was required.

The three basic components necessary for a targeting system are: (1) a target present either exclusively on the tumor cell or at least expressed in greater amounts on tumor compared to normal tissue; (2) a carrier or delivery system; (3) a toxic agent or set of molecules which will cause damage to the tumor after they have been directed there. The objective with this form of targeted therapy (site-specific delivery of drugs or toxins) is to deliver the toxic agent to its site of action on or near the tumor cells, thereby reducing toxicity to normal cells and increasing the therapeutic index of the agent.

Ideally, the target would be present on the tumor and not on any normal cells or tissues. Unfortunately, despite extensive investigation, the existence of tumor-specific antigens remains unproven for the majority of human tumors (but see also Section 12.6.1) and most antibody-mediated targeting has therefore been directed at tumor-associated antigens, many of which are differentiation antigens such as carcinoembryonic antigen (CEA), or antigens present on particular cell types such as the melanoma-associated antigens. A variety of other cancer markers have also been investigated. Even if no truly tumor-specific targets are available, products of oncogenes that are qualitatively different from normal host products could serve to increase the therapeutic index.

12.4   Use of antisense RNA or oligonucleotides

Following the initial discoveries of natural antisense RNAs in prokaryotes, numerous applications of antisense RNA-mediated regulation have been demonstrated in a variety of experimental systems [31, 32]. These nontranslated mRNAs directly repress gene expression by hybridizing to a target RNA, rendering it functionally inactive. Specificity of antisense RNA for a particular transcript is conferred by extensive sequence complementarity with the 'sense' or target RNA. Translation of a target mRNA is inhibited following formation of a sense–antisense RNA hybrid. In addition, the duplex molecule may become sensitive to double-strand-specific cellular nucleases. Other effects of antisense RNA may include transcriptional attenuation of the mRNA and also disruption of post-transcriptional processing events [32].

Oncogene DNA and RNA differ in nucleotide sequences from normal proto-oncogene DNA and RNA, and it is therefore theoretically possible to design specific antisense molecules to block translation of oncogene mRNA. For example, by analogy with the microinjection of antibodies against the RAS p21 oncogenic product cited earlier, microinjection of antisense DNA sequences might bind to, and block expression of, oncogene DNA. Such 'blocking DNA' could also incorporate an anti-DNA toxin, such as an alkylating agent, to destroy the portion of the chromosome encoding the oncogene specifically. Similarly, an antibody against the oncogenic RNA sequence might bind and prevent that mRNA molecule from being used by ribosomes, or it might inhibit elongation by blocking transfer RNA from binding to the oncogenic messenger anti-codon [2].

Several groups have tried to reverse the transformed phenotype by expressing large amounts of mRNA from the DNA strands complementary to the one coding an aberrant oncogene protein. In the nuclei of the cells the two complementary mRNA strands hybridize to form a double-stranded structure that effectively prevents translation of the mRNA [2] (*Figure 12.5*). It is now possible to design antisense DNA oligonucleotides, or catalytic antisense RNAs (ribozymes) which can pair with and functionally inhibit the expression of any single-stranded nucleic acid in a specific fashion. This high degree of specificity has made them attractive candidates as therapeutic agents [33]. These oligonucleotides demonstrated important early application to the elucidation of cellular signaling pathways. More recently, studies with these agents have probed their utility as potential therapeutic agents, especially in the realm of cancer. With the implementation of gene therapy in early clinical trials, oligonucleotide mediated suppression of gene expression has emerged as an important strategy for gene therapy [34, 35].

Several laboratories are testing retroviruses encoding antisense RNA expressed in a tissue-specific manner by use of appropriate enhancer and other control regions attached to the anticancer genes. To make this approach practical, a better understanding of retroviral tissue tropism mediated by cell surface receptors will be important. It would not be acceptable to infect a cancer patient

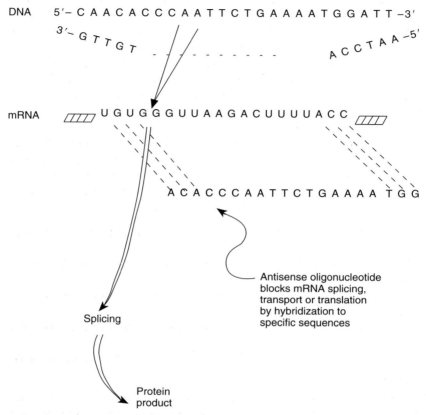

Figure 12.5: Principles of antisense therapeutics. Oligonucleotides or retroviral encoded antisense mRNAs bind to critical control regions (e.g. splice acceptor sites) and prevent the processing, extranuclear transport and translation of selected transcripts. Reproduced from Buick *et al.* (1988) in *Oncogenes: an Introduction to the Concept of Cancer Genes*, Springer-Verlag, with permission from Springer-Verlag.

with a retroviral vector carrying an antisense gene until its efficacy and more importantly its safety have been proved [2]. Despite this caveat, such an approach is already being used clinically (see Section 12.7).

Attempts to realize the promise and expectations raised by the concept of antisense blocking of specific genes involved in cancer, AIDS and a variety of other diseases have resulted in some unexpected questions arising about how these genes really work [36, 37]. Even though the phosphorothioates are generally believed to represent the 'first generation' of antisense nucleotides, they suffer from certain drawbacks like the existence of diastereoisomeric mixtures, intermediate RNA binding affinity and certain nonspecific side effects [38]. Solutions for obtaining antisense oligonucleotides with improved binding properties and nuclease resistance can be realized by various chemical means

[39]. In addition, novel approaches which can influence the biological efficacy by modulating the biodistribution and/or stability of the target mRNA by cleavage have been proposed. However, whether these new promising candidates will be useful as a 'second generation' of antisense compounds remains to be proven by their *in vivo* pharmacokinetic and pharmodynamic behavior. So far, the collection of *in vivo* data is limited exclusively to methylphosphonates and in particular phosphorothioates which have entered clinical trials as the 'first generation' of antisense compounds [39].

As stressed in a recent review of the antisense treatment of viral infection [40] many simple but critical questions remain unanswered and this is also true of its application in cancer. A current review [41] focuses on those aspects of chemistry and mechanism that are thought to be important and relevant for the therapeutic use of the current generation of deoxynucleotide agents. These authors designate the outstanding questions as being:

(1)   how can oligonucleotides be delivered more efficiently to intracellular compartments?
(2)   How can oligonucleotides be more selectively delivered from the bloodstream to target tissues?
(3)   How stable must an oligonucleotide be for optimal function and low toxicity?
(4)   Will development of new RNase H-independent inhibitors lead to greater specificity?
(5)   Can new agents be developed that avoid aptameric effects and inhibit by a purely antisense mechanism?

The first annual *Nature Medicine* conference, 'The Art of Antisense', was held in September, 1995, in New Orleans. Readers are referred to the reviews cited earlier [40, 41] as well as to the commentaries by the conference organizers on the current data, and on the promise and shortcomings [42, 43], to gain an up-to-date perspective on this important and promising field of antisense.

12.5   Cytokine modulation of oncogenes

In addition to the targets represented by the protein products of oncogenes, there is evidence that the expression of mutant genes is closely linked to aberrant cytokine gene expression. It has been demonstrated that expression of a mutant *RAS* oncogene in at least two different human cell types is associated with significant alterations in the regulation of genes encoding several cytokines, including IL-1A, IL-1B, CSF2, CSF3 and IL-6 [44]. Further elucidation of the mechanism by which *RAS* mutations result in dysregulation of cytokine gene expression should lead to a better understanding of the biological effects of these genetic alternatives in cancer cells, and will possibly reveal new ways of manipulating the growth of malignant cells.

It has recently been demonstrated experimentally in a mouse P815 mastocytoma model that tumor cells, transfected with a human IL-13 cDNA in a plasmid expression vector, when injected into mice were rejected and that the mice developed systemic specific anti-tumor immunity leading to long-lasting specific anti-tumor protection [45]. In another study in mice, the vascular delivery of retroviral producer cell lines encoding the cytokines IL-4 and IL-2, directed tumoricidal inflammatory responses to established metastases. Cytokine gene targeting inhibited metastasis formation and caused significant overall reduction in tumor burden [46], indicating the potency of this approach.

Some growth factors are therapeutically useful partly because restricted expression of their receptors limits their action to particular cell types. However, no unique stimulating factor is known for many clinically relevant cell types. Recent studies have shown that it is possible to target the soluble α receptor (Rα) component for IL-6 and ciliary neurotrophic factor (CNTF) to cells expressing particular surface markers, thereby rendering these cells responsive to IL-6 or to CNTF. This approach may open the way for 'designer cytokines' that are tailored to stimulate cells of choice, especially those for which a unique cytokine may not exist [47].

12.6   Tumor suppressor genes as targets

The therapeutic possibilities of introducing extra copies of the wild-type *p53* and *RB1* genes into the DNA of tumor cells are readily apparent because the loss of normal function of one or both of these gene products has proved to be a frequent event in the development of human cancer. In the case of both *RB1* and *p53*, returning the wild-type allele of these genes into a transformed cell, or a cell undergoing transformation in culture, reverses the tumorigenic potential of the cell or prevents transformation occurring [48]. There is now convincing evidence that both *RB1* and *p53* have the ability to reverse the malignant phenotype when wild-type genes and gene products are expressed in transformed cells. In future it may prove possible to provide extra copies of these tumor suppressor genes, or others that may be identified subsequently, prophylactically to all somatic cells, thus reducing the risk of developing cancer. However, 'simple' gene insertion may not be enough as it has been shown that different amounts of the *RB1* gene product, RB, are required to counter the susceptibility to cancer than to prevent other conditions, such as developmental abnormalities [49], and control of expression of the protein product of the replacement tumor suppressor gene may prove to be crucial.

12.6.1   p53 *as a target for therapy*

The possibility of new therapeutic approaches is also raised by the fact that p53 protein is so commonly mutated, revealing new epitopes, that a tumor-specific *p53* antigen has been demonstrated in the majority of human cancers (see also Section 12.2.1). Over the 15 years since it was discovered, *p53* has progressed from being a curiosity involved in SV40 viral transformation to being the site of the most common genetic change in human cancer and has thereby opened up new therapeutic approaches [50]. As well as the possibilities indicated in the previous section, it may also be possible to immunize patients appropriately with p53 protein and to stimulate a cytotoxic T-cell response against their tumors.

The *p53* tumor suppressor gene can prevent the development of cancer by blocking the division of cells that have sustained DNA damage as well as by triggering apoptosis. This function of *p53* must often be rate limiting as inactivation of *p53* is one of the most common molecular events in the development of cancer [51]. There is much that is still not known about *p53* and several questions which need to be answered have been posed recently [51]. As suggested by this author, answers to these questions should offer new ideas and routes towards improved prevention, diagnosis and treatment of cancer.

Most *p53* mutants are either DNA contact mutants (which are unable to bind specific target sequences because of substitution of crucial DNA contact residues, including mutational 'hot-spots') or structural mutants which lose specific DNA binding capacity due to abnormal conformation of the protein [52]. These two classes of mutant may prove amenable to different anticancer therapies designed to restore wild-type tumor suppressor function. These include: reintroducing a wild-type *p53* gene using adenoviral and retroviral replacement vectors or liposomal gene insertion for gene therapy; restoring wild-type *p53* function by refolding the protein to its normal conformation; use of phamacologic agents which mimic or restore *p53* function; inducing immunological recognition of structural mutant-specific peptides associated with the major histocompatibility complex (MHC) molecules on the tumor cell surface [52, 53].

Programed cell death, or apoptosis, may be triggered by stimulation of *p53*. Whether or not *p53* causes reversible growth and/or apoptosis depends partly on the state of cellular activation. Conflicting growth regulatory signals received during *p53* induction, sustained *p53* synthesis because of extensive unrepaired DNA damage, or *p53* activation after irreversible commitment to replication all drive a cell to apoptosis [54].

There are *p53* dependent and *p53* independent pathways to apoptosis [55]. However, once the precise biochemical links between *p53* activation and apoptosis are identified it may be possible to develop pharmacological agents that mimic their actions and thereby provide another approach to treating cancer more effectively [53].

Some scientific discoveries enjoy a meteoric rise and then a meteoric fall, *p53* is an exception, as it has helped to shed light on molecular carcinogenesis over the last 16 years although no therapeutic benefit has yet been realized from this knowledge [53]. However, the development of successful strategies for gene replacement or pharmacological restoration of *p53* function holds the promise of significant therapeutic advances in the future.

12.7   Gene therapy

The potential for gene therapy of oncogenes has been alluded to several times in the preceding pages. For an erudite discussion of the problems associated with gene therapy in humans, readers are referred to Weatherall's excellent book [56]. Although this book was published several years ago the issues he raises are as relevant today as they were then. Up until now, the twin Achilles heels of gene therapy have been the inability to target specifically and the inability to control exogenous gene expression [53].

An example of a successful approach in an animal model is the liposomal delivery of the human *APC* gene to rodent colonic epithelium where 100% of the epithelial cells which contacted the liposome–gene mixture were shown to have taken up the gene [57]. Expression was, however, only transient and did not persist beyond 3 days, thereby limiting the potential therapeutic utility of a single dose. Permanent expression of genes in colonic epithelium is not feasible with current technology, however, continuous expression can be obtained by repeated treatment with liposomal enemas every 2–3 days. If studies currently underway in rodents indicate that this will result in the inhibition or reversal of growth of colonic cancers, then clinical trials of this approach in FAP patients would be warranted [57].

As studies such as this in animal models have begun to reveal, major impediments remain to be overcome if gene therapy is to become an everyday reality. One of these is that adenoviral vector administration commonly induces inflammation and antigen-specific cellular and humoral immune responses [58]. Immunomodulation, inducing transitory immunosuppression, by using recombinant IL-12 to block the production of IgA antibodies is a recent development towards overcoming this problem [59].

Despite these problems a number of clinical trials of gene therapy for the treatment of cancer have been approved or are already underway. One of the most promising approaches has been the use of retroviral vectors containing a herpes simplex virus thymidine kinase (HSVtk) gene in patients with brain tumors. In rat models it had previously been shown that it was mainly the rapidly proliferating tumor cells that integrated HSVtk and subsequent intravenous infusions with gancyclovir resulted in considerable death of cells expressing HSVtk [60 and references therein]. Results of clinical studies using this approach are awaited with great interest.

However, most research continues to be in the area of vector development. Crucial to this is the development of efficient, multipurpose therapeutic gene delivery systems which should have the following characteristics:

(1)   the ability to target specific cells;
(2)   no limitation on the size or type of nucleic acid that can be delivered;
(3)   no intact viral component and therefore be safe for the recipient;
(4)   the ability to transduce a large number of cells regardless of their mitotic status;

(5)   the potential to be completely synthetic.

The development of molecular conjugates (ligands to which a nucleic acid or DNA-binding agent has been attached for the specific targeting of nucleic acids, i.e. plasmid DNA, to cells as delivery vectors) has resulted in the creation of a simple, nonviral method for the targeted delivery of nucleic acids to specific cell types. These 'synthetic viruses' can easily incorporate components that are associated with viral function [61 and references therein].

12.8   Conclusions and future prospects

Much of the work on oncogene therapeutics is in a preliminary or even speculative theoretical stage. There is considerable enthusiasm for the therapeutic potential revealed by the study of oncogenes; however, it must be remembered that oncogene products are of fundamental importance to normal cell growth and differentiation, and that mutant oncogenes and their products often differ from their normal counterparts only in minor details. In theory this may be sufficient to improve the current therapeutic index, but it is possible that anti-oncogene therapies may prove equally, or more, toxic than the therapeutic modalities currently available. With few exceptions, the technical means available to achieve these therapeutic possibilities still range from the primitive and experimental to the currently impossible, making practical realization of many of these treatments remain a good number of years in the future [2]. However, as indicated above, progress towards developing effective therapies is being made.

In addition, the demonstration of anti-p53 antibodies in the sera of patients with a variety of cancers [62, 63 and references therein], the *in vitro* induction of cytotoxic T cells specific for wild-type and mutant p53 peptides [64] as well as the demonstration that the immunization of patients with a vaccinia virus vector containing a CEA peptide can induce cytotoxic T cells which recognize MHC restricted binding motifs in CEA [65], indicate that it is possible to generate immune responses to the 'self antigens' expressed by tumor cells and offers an additional potential avenue for therapeutic intervention in the future.

References

1.   Glover, D.M. and Hames, B.D. (1989) *Oncogenes*. IRL Press, Oxford.
2.   Buick, K.B., Liu, E.T. and Larrick, J.W. (1988) in *Oncogenes: an Introduction to the Concept of Cancer Genes*. Springer-Verlag, New York, p. 262.
3.   Kohler, G. and Milstein, C. (1975) *Nature*, **256**, 495.
4.   Ford, C.H.J. and Casson, A.G. (1986) *Cancer Chemother. Pharmacol.*, **17**, 197.
5.   Vitetta, E.S., Thorpe, P.E. and Uhr, J.W. (1993) *Immunol. Today*, **14**, 252.
6.   Wels, W., Moritz, D., Schmidt, M., Jeschke, M., Hynes, N.E. and Groner, B. (1995) *Gene*, **159**, 73.
7.   Pastan, I., Pai, L.H., Brinkman, U. and FitzGerald, D. (1996) *Breast Cancer Res.*

Treat., **38**, 3.
8.   Colnaghi, M., Ménard, S. and Canevari, S. (1993) *Curr. Opin. Oncol.*, **5**, 1035.
9.   Pietersz, G.A., Rowland, A., Smyth, M.J. and McKenzie, I.F.C. (1994) *Adv. Immunol.*, **56**, 301.
10. Goldenberg, D.M., Larson, S.M., Reisfeld, R.A. and Schlom, J. (1995) *Immunol. Today*, **16**, 261.
11. Frankel, A.E., FitzGerald, D., Siegall, C. and Press, O.W. (1996) *Cancer Res.*, **56**, 926.
12. Wawrzynczak, E.J. (Ed.) (1995) *Antibody Therapy*. BIOS Scientific Publishers, Oxford.
13. Williams, A.F. and Barclay, A.N. (1988) *Ann. Rev. Immunol.*, **6**, 381.
14. Lewis, (1994) *Adv. Immunol.*, **56**, 27.
15. Adair, J.R. (1992) *Immunol. Rev.*, **130**, 5.
16. Shin, S-U., Wright, A., Bonagura, V. and Morrison, S.L. (1992) *Immunol. Rev.*, **130**, 87.
17. Neuberger, M.S., Williams, G. and Fox, R.O. (1984) *Nature*, **312**, 604.
18. Bruggemann, M., Williams, G.T., Bindon, C., Clark, M.R., Walker, M.R., Jefferis, R., Waldmann, H. and Neuberger, M.S. (1987) *J. Exp. Med.*, **166**, 1351.
19. Bruggeman, M., Winter, G., Waldman, H. and Neuberger, M. S. (1989) *J. Exp. Med.*, **170**, 2153.
20. Jones, P.T., Dear, P.H., Foote, J., Neuberger, M.S. and Winter, G. (1986) *Nature*, **314**, 268.
21. Riechmann, L., Clark, M., Waldmann, H. and Winter, G. (1988) *Nature*, **332**, 323.
22. McCafferty, J., Griffiths, A.D., Winter, G. and Chiswell, D.J. (1990) *Nature*, **348**, 552.
23. Soderlind, E., Simonsson, A.C. and Borrebaek, C.A.K. (1992) *Immunol. Rev.*, **130**, 109.
24. Winter, G., Griffiths, A.D., Hawkins, R.E. and Hoogenboom, H.R. (1994) *Ann. Rev. Immunol.*, **12**, 433.
25. De Kruif, J., Terstapper, L., Boel, E. and Logterbug, T. (1995) *Proc. Natl Acad. Sci. USA*, **92**, 3938.
26. Feramisco, J.R., Clark, R., Wong, G., Arnheim, N., Milley, R. and McCormick, F. (1985) *Nature,* **314**, 639.
27. Biocca, S. and Cattaneo, A. (1995) *Trends Cell Biol.*, **5**, 248.
28. Perez, P., Titus, J.A., Lotze, M.T., Cuttitta, F., Longo, D.L., Groves, E.S., Rabin, H., Durda, P.J. and Segal, D.M. (1986) *J. Immunol.*, **137**, 2069.
29. Baserga, R. (1994) *Cell*, **79**, 927.
30. Rodeck, U., Williams, N., Murthy, U. and Herlyn, M. (1990) *J. Cell. Biochem.*, **44**, 69.
31. Green, P.J., Pines, O. and Inouye, M. (1986) *Ann. Rev. Biochem.*, **55**, 569.
32. Takayama, K.M. and Inouye, M. (1990) *Crit. Rev. Biochem. Mol. Biol.*, **25**, 155.
33. Rossi, J.J. (1995) *Br. Med. Bull.*, **51**, 217.
34. Scanlon, K.J., Ohta, Y., Ishida, H., Kijima, H., Ohkawa, T., Kaminski, A., Tsai, J., Horng, G. and Kashani-Sabet, M. (1995) *FASEB J.*, **9**, 1288.
35. Putnam, D.A, (1996) *Am. J. Health-Syst. Pharm.*, **53**, 151.
36. Gura,T. (1995) *Science*, **270**, 575.
37. Rojanasakul, Y. (1996) *Adv. Drug Deliv. Rev.* **18**, 115.
38. Stein, C.A. and Cheng, Y.C. (1993) *Science*, **261**, 1004.
39. De Mesmaeker, A., Haner, R., Martin, P. and Moser, H.E. (1995) *Acc. Chem. Res.*, **28**, 366.

40. Whitton, J.L. (1994) *Adv. Virus Res.*, **44**, 267.
41. Heidenreich, O., Kang, S.-H., Xu, X. and Nerenberg, M. (1995) *Molec. Med. Today*, **1**, 128.
42. Wagener, R. (1995) *Nature Med.*, **1**, 1116.
43. Stein, C.A. (1995) *Nature Med.*, **1**, 1119.
44. Demetri, G.D., Ernst, T.J., Pratt, E.S. II., Zenzie, B.W., Rheinwald, J.G. and Griffin, J.D. (1990) *J. Clin. Invest.*, **86**, 1261.
45. Lebel-Binay, S., Laguerre, B., Quintin-Colonna, F., Conjeaud, H., Magazin, M., Miloux, B., Pecceu, D., Ferrara, P. and Fradelizi, D. (1995) *Eur. J. Immunol.*, **25**, 2340.
46. Hurford, R.K., Dranoff, G., Mulligan, R.C. and Tepper, R.I. (1995) *Nature Genetics*, **10**, 430.
47. Economides, A.N., Ravetch, J.V., Yancopoulos, G.D. and Stahl, N. (1995) *Science*, **270**, 1351.
48. Levine, A.J. and Momand, J. (1990) *Biochim. Biophys. Acta*, **1032**, 119.
49. Skuse, G.R. and Ludlow, J.W. (1995) *Lancet*, **345**, 902.
50. Harris, A.L. (1990) *J. Pathol.*, **162**, 5.
51. Lane, D. (1994) *Int. J. Cancer*, **57**, 623.
52. Milner, J. (1995) *Nature Med.*, **1**, 879.
53. Elledge, R.M. and Lee, W.-H. (1995) *BioEssays*, **17**, 923.
54. Carson, D. and Lois, A. (1995) *Lancet*, **346**, 1009.
55. Clarke, A.R., Purdie, C.A., Harrison, D.J., Morris, R.G., Bird, C.C., Hooper, M.L. and Wyllie, A.H. (1993) *Nature*, **362**, 849.
56. Weatherall, D.J. (1991) *The New Genetics and Clinical Practice*, 3rd Edn. Oxford University Press, Oxford.
57. Westbrook, C.A. and Arenas, R.B. (1995) *Adv. Drug Deliv. Rev.*, **17**, 349.
58. Wilson, C. and Kay, M.A. (1995) *Nature Med.*, **1**, 887.
59. Yang, Y., Trinchieri, G. and Wilson, J.M. (1995) *Nature Med.*, **1**, 890.
60. Culver, K.W. and Blaese, R.M. (1994) *Trends Genet.*, **10**, 174.
61. Cristiano, R.J. and Roth, J.A. (1995) *J. Mol. Med.*, **73**, 479.
62. Houbiers, J.G.A., van der Burg, S.H., van de Watering. L.M.G., Tollenaar, R.A.E.M., Brand, A., van de Velde, C.J.H. and Melief, C.J.M. (1995) *Br. J. Cancer*, **72**, 637.
63. Trivers, G.E., Cawley, H.L., DeBenedetti, V.M.G. *et al.* (1995) *J. Natl Cancer Inst.*, **87**, 1400.
64. Houbiers, J.G.A., Nijman, H.W., van der Burgh, S.H. *et al.* (1993) *Eur. J. Immunol.*, **23**, 2072.
65. Schlom, J., Kantor, J., Abrams, S., Tsang, K.Y., Panicali, D. and Hamilton, J.M. (1996) *Breast Cancer Res. Treat.*, **38**, 27.

Chapter 13

Molecular techniques for analysis of genes

13.1   Introduction

An understanding of the methodology used in the analysis of genes described in earlier chapters is fundamental to understanding how these genes can be used for the diagnosis and prognosis of cancer. There are many excellent molecular biology books available which describe these procedures in detail [1–3]. This chapter outlines these techniques briefly so that those unfamiliar with the technology can follow the previous chapters without immediate recourse to other books.

The genes can be studied at three levels, namely DNA, RNA or protein. These molecules can be either within the cell in prepared tissue sections (*in situ*) or in isolation. These two approaches give different information. Isolated DNA can be examined for qualitative and quantitative abnormalities. This is useful when looking for rearrangements or mutations within a particular gene and also for assessing absolute levels of a gene (gene amplification). Analysis of isolated RNA gives information about the level of transcription (gene expression). Direct analysis of protein allows the determination of protein levels or of changes in protein size.

In situ analysis provides information concerning the spatial distribution of molecules in the cell and therefore shows which cells are expressing a particular gene, RNA sequence or protein molecule, but cannot easily be used for quantitative analysis. When used on chromosome spreads this technique gives information about the chromosomal localization of a gene or, once a gene has been mapped, can be used to identify gene deletions.

13.2   Analysis of DNA or RNA

13.2.1   Genetic markers

Three different types of DNA marker are used in genetic analysis. The markers segregate in Mendelian fashion and, provided they are sufficiently polymorphic, can be used in gene tracking studies, to help make genetic maps to be used in the subsequent isolation of genes and in cancer research for the identification of loss of heterozygosity (LOH).

The first markers to be developed were the restriction fragment length polymorphisms (RFLPs). Approximately every 200 bp along the length of the chromosomes the sequence of DNA varies between individuals, usually by only a single base change. Where they coincide with restriction sites, the alterations can be detected by restriction enzymes. The presence or absence of any particular site is variable in the population, hence the term polymorphism, and the technique is referred to as RFLP analysis. The variations do not usually confer any phenotypic effect since they are frequently found in introns rather than coding exons (*Figure 13.1a*). Approximately 100 000 RFLPs are distributed throughout the human genome.

The second group of markers is composed of arrays of short tandem repeated sequences, each different allele being made up of a different number of repeats (*Figure 13.1b*). These markers are termed minisatellites or variable number of tandem repeats (VNTRs) and approximately 10 000 exist in the genome [4]. Many of these markers are located towards the telomeres of the chromosomes which can limit their usefulness.

The third group, which are the ones now most frequently in use for genetic analysis, are the microsatellites [5]. The most common of these markers are the CA repeats which are composed, as the name suggests, of repeats of the CA dinucleotide (*Figure 13.1c*). However other microsatellites, primarily tri- and tetranucleotide markers, are also available and have some technical advantages over dinucleotide repeats. These markers are found in the genome with approximately the same frequency as the VNTRs.

The reasons why the latter two groups of markers have found favor over RFLPs is primarily because they are much more informative for genetic studies, as discussed below (see Section 13.2.5). Microsatellites have the added advantage of being analyzable by PCR, a technique which revolutionized much of molecular biology and which is described in the next section. The first two groups of markers are studied using Southern blotting techniques (see Section 13.2.4) and therefore take much longer to analyze.

13.2.2 Polymerase chain reaction (PCR)

In 1985, a technique allowing specific amplification of selected sequences of DNA was described. PCR technique relies on a knowledge of at least part of the DNA sequence around the region of interest, because short specific oligonucleotides complementary to sequences either side of this region are required to prime the synthesis of the DNA sequence between them. The process involves three stages: denaturation of the double-stranded DNA, annealing of the oligonucleotide primers, and synthesis of DNA by a thermostable DNA polymerase. By repeating the process around 30 times, up to several million copies of the DNA can be made (*Figure 13.2*). The main advantages of this technique are: (1) it is rapid, particularly as the product can often be detected directly on agarose gels (this takes several hours to perform, unlike

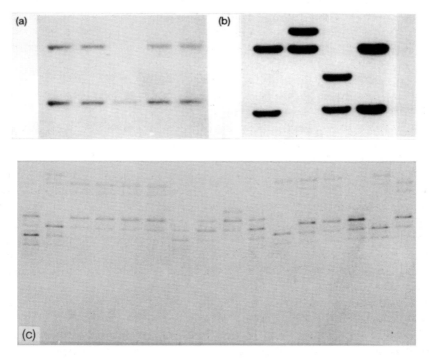

Figure 13.1: (a)   Two allele RFLP linked to the APC gene. (b) Minisatellite marker on chromosome 16p showing several different alleles present in the population. (c) Microsatellite analysis for the marker LNS linked to the *APC* gene. One PCR primer has been end labeled with ^{35}S, the products analyzed on a 6% denaturing gel and exposed to X-ray film. Courtesy of Steve McKay, DNA Laboratory, Regional Genetics Service, Birmingham Heartlands Hospital.

Southern blotting which can take several days or more); (2) it is extremely sensitive (it is possible to start with a single molecule of DNA and finish with sufficient for analysis [6]); (3) DNA from unusual sources can be used (see Section 13.2.5).

PCR forms the basis of many analyses. The most important of these, with reference to the areas covered in this book, are the typing of microsatellite markers either for gene tracking studies as discussed below (see Section 13.2.5) or for the determination of loss of heterozygosity (see Section 3.2.2), for the detection of point mutations as discussed in the next section, or for the production of template DNA for sequencing.

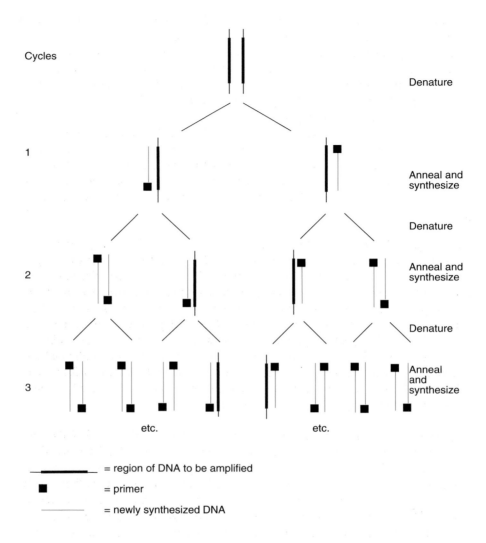

Figure 13.2:   Polymerase chain reaction.

13.2.3 Mutation detection techniques

Now that many genes have been isolated, the direct analysis of mutations has become routine practice. This procedure falls into two main groups: (1) some techniques are used where only one or two mutations are commonly found to disrupt the gene for example the common mutations in the *RAS* gene; (2) other techniques are used as a preliminary screen to identify approximately where in a gene a mutation lies. This can then be followed up by sequencing of that region to confirm the nature of the mutation. These techniques have been used extensively in identifying mutations in the genes causing inherited cancers. In all of these genes described so far, mutations not only vary from one family to the next but in the majority of cases are relatively small abnormalities such as point mutations or small deletions and insertions and are widely distributed throughout the gene. Each of these techniques described below has different advantages such as sensitivity, ease of the procedure, cost and safety, so individuals have to balance each of these when selecting the method to use.

All of the procedures described below depend on initial PCR amplification of a region of the gene, usually the coding sequence and often a single exon, followed by analysis on a variety of different gels. Some of the more commonly used techniques are described briefly below and further details can be found in reference [7].

Allele-specific oligonucleotide (ASO) hybridization.   Where only one or two common mutations, differing from the normal sequence by as little as a single base change, are known to disrupt a gene, ASOs can be used to identify the presence of the mutation. This technique involves the hybridization of the PCR product, bound to membrane filters in a denatured state, with oligonucleotide probes specific for a known mutation. These oligonucleotide probes are hybridized under specific hybridization conditions such that they only bind to the target DNA if there is perfect complementarity between the probe and the target. This technique has been used successfully to analyze the distribution of *RAS* mutations at codons 12, 13 or 61 in different tissues as described in reference [8] and in many of the preceding chapters.

Single-stranded conformation polymorphism analysis (SSCP).   SSCP is one of the most commonly used techniques for mutation detection because of its simplicity and relatively good sensitivity. Wild-type and test DNA are amplified by PCR and then denatured. The single-stranded molecules from each PCR product adopt a three-dimensional structure dependent on the sequence of the DNA. They are electophoresed through a nondenaturing gel and the migration of the fragment is dependent on the conformation adopted. If a mutation is present there will be a difference in the migration rate. Bands are visualized either by incorporation of a radioisotope during the PCR reaction or by post electrophoretic staining of the gel. The presence of a variation in the sequence is detected by comparison of the test sequence with a known wild-type fragment (*Figure 13.3*). The sensitivity of SSCP is around 70–90% for PCR fragments less than 200 bp in length but decreases rapidly if fragments above 400 bp are analyzed. SSCP analysis can also be carried out on Hydrolink or MDE™ gels with a similar detection rate (see *Figure 6.7*).

Heteroduplex analysis.   This technique depends on the fact that if both wild-type and mutant DNA are present in a PCR reaction, heteroduplex molecules are formed between the two different DNA species during the final rounds of amplification. Heteroduplex molecules migrate through acrylamide gels differently to homoduplex molecules, a feature which is enhanced by the use of gel matrices such as Hydrolink or MDE™ gels. The sensitivity of this technique is similar to SSCP analysis with the advantage of being extremely simple to perform. The sensitivity of the technique decreases if fragments over 300 bp are analyzed.

Denaturing gradient gel electrophoresis analysis (DGGE).   DGGE relies on the fact that a double-stranded DNA fragment exposed to a denaturant will melt in discrete domains to become single stranded and will do so in a sequence-specific manner. A single base change in the DNA is sufficient to change its melting temperature. In this technique, double-stranded DNA fragments are electophoresed through a gel containing an increasing concentration of denaturant. As the DNA migrates, individual domains of the fragment melt at a point corresponding to their melting temperature. As the fragment becomes partially melted, its migration through the gel is retarded. In order to prevent complete

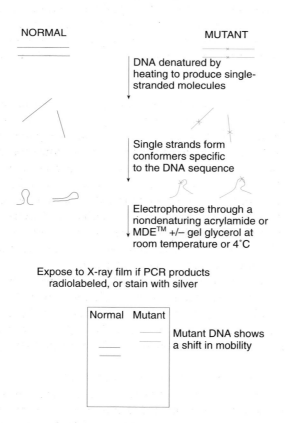

Figure 13.3:   SSCP analysis.

dissociation of the fragment, it is necessary to add a GC clamp (a sequence of approximately 40 bases, composed primarily of Cs and Gs) to one PCR primer; this region has a very high melting temperature so it will ensure that the fragment never separates into purely single-stranded molecules which would have no variation in their migration. The procedure is further enhanced by the production of heteroduplex molecules formed between one mutant and one wild-type DNA strand during the final rounds of PCR. These heteroduplexes have different melting temperatures compared with either wild-type or mutant homoduplexes and can sometimes be detected more readily on the gels (see *Figure 6.5*). Detection of DNA fragments can be either by isotopic or nonisotopic means. DGGE is one of the most sensitive techniques as it can detect close to 100% of mutations in fragments up to 600 bp. However it is technically more difficult to set up initially and is also more expensive. The denaturing gradient can also be replaced with a temperature gradient with similar results.

DGGE analysis can also be carried out in two dimensions. For a patient under investigation, each exon of a gene is amplified and products are separated in the first dimension, based on size, on a neutral gel. The track containing the fragments is then cut out of the gel, turned through 90° and layered across a denaturing gradient and electrophoresed in the second dimension. Again any abnormal fragments will migrate differently from wild-type ones. The advantage of this technique is that a whole gene can be analyzed for a single individual in one gel.

Chemical mismatch cleavage (CMC) and RNase cleavage. In this technique heteroduplexes are formed between radiolabeled wild-type DNA and mutant DNA by denaturing and reannealing. Any mismatches are chemically modified by osmium tetroxide for mismatched thymines or by hydroxylamine for mismatched cytosines. Mismatched guanines and adenines are also detected by labeling the antisense strand. The sites of chemically modified mismatches are then cleaved with piperidine, the products electrophoresed on denaturing acrylamide gels and detected by autoradiography (*Figure 13.4*). This technique is at least as sensitive as DGGE and also has the advantage that the position of the mutation can be localized. However its main drawbacks are the hazardous nature of the chemicals used and also that it is labor intensive.

Another technique based on the same principle is RNase A cleavage. In this method a heteroduplex molecule is formed between a wild-type riboprobe and a DNA fragment from the patient in question. Any mismatches formed because of sequence differences due to mutations can be cleaved by RNase. The main problem with this technique is its relatively low sensitivity of around 40–60%

RT-PCR. If the sequences of intron/exon boundaries are not known it is not possible to design PCR primers either side of the exon, so that it cannot be analyzed by the above techniques. One way around this problem is to produce cDNA from mRNA by reverse transcriptase PCR (RT-PCR). Primers lying within the coding sequence can then be used to amplify regions of the gene for further analysis. The technique can however only be used where the gene is expressed in accessible tissues. The analysis of cDNA is helpful for the identification both of larger deletions and of mutations affecting the splicing of mRNA which give smaller products because of the loss of several exons. In general, these differences can be identified by analysis of the cDNA from normal and affected individuals on agarose gel electrophoresis simply by the presence of small fragments in affected individuals. cDNA can also be the basis for the protein truncation test (PTT), described below, as large sections of coding sequence can be analyzed in a single reaction. For example, exons 1–14 of the *APC* gene can be analyzed together rather than in 14 separate reactions as used for SSCP or DGGE. Similarly, cDNA is useful in techniques such as chemical mismatch cleavage in which large fragments (1–2 kb) can be analyzed successfully thereby reducing the amount of screening necessary to cover a gene.

Protein truncation test (PTT).   Many of the causative mutations in cancer genes are chain-terminating. A novel approach to mutation detection which makes use of this fact is the PTT, originally described for detection of point mutations in Duchenne muscular dystrophy [9]. The basis of this technique is again PCR but here the primers are modified to include the T7 promoter sequence as well as a eukaryotic initiation signal plus the usual sequence-specific region. Following PCR amplification of either cDNA or genomic DNA (if an exon is sufficiently large such as exon 15 of *APC* or exon 11 of *BRCA1*),

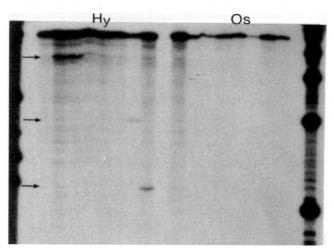

Figure 13.4:   Chemical mismatch cleavage. Two groups of eight samples have been treated either with hydroxylamine (Hy) or osmium tetroxide (Os). The arrows indicate bands in two patients with similar mutations in lanes 2 and 3. Two other mismatches can be seen in lanes 7 and 8. All of these bands are seen with hydroxylamine only, indicating that they involve a cytosine. Molecular weight markers are included in the extreme right- and left-hand lanes. Photograph courtesy of Dr S. Enyat, DNA and Regional Haemostasis Laboratory, Birmingham Children's Hospital.

the PCR products are used as templates in transcription and translation reactions resulting in the production of a peptide from the PCR fragment. In order to detect the full length or truncated peptide, a radiolabeled amino acid is included in the translation reaction. The products of the translation reaction are separated on an SDS–polyacrylamide gel and visualized by autoradiography (*Figure 13.5*). The advantage of this technique is that whole genes can be analyzed in a few reactions each covering several kilobases [10]. However for the majority of the genes, RNA has to be used as starting material and cDNA made as the initial template for the PCR reaction. The major drawback to the technique is that it can only pick up truncating mutations. However it could in theory be adapted to use isoelectric focusing as the detection step when it could presumably also pick up missense mutations.

Direct sequencing. The gold standard for mutation detection is direct sequencing of the PCR fragment. This will immediately give information about the nature of the mutation and any changes due to neutral polymorphisms can be identified. Sequencing is usually based on the Sanger dideoxy chain-termination method and is traditionally carried out radioactively. Many kits are now available for cycle sequencing based on this method which makes use of PCR and means that only minute amounts of template are required. Fluorescence-based sequencing methods are becoming increasingly widespread. These are based either on cycle sequencing or on Sequenase™ based methods and use either fluorescently labeled primers or fluorescently labeled dideoxy terminators. These methods are the most efficient since they can be

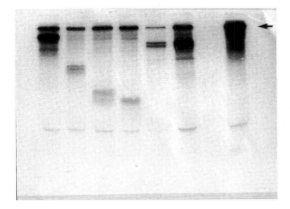

Figure 13.5: PTT for exon 15 (segments A to F) of the *APC* gene. Truncated peptides are seen in lanes 1–6. The full length transcript corresponding to the product of the normal gene is arrowed. Courtesy of Steve McKay, DNA Laboratory, Regional Genetics Service, Birmingham Heartlands Hospital.

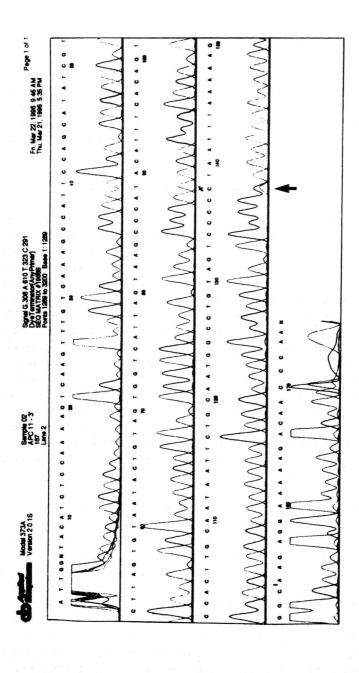

Figure 13.6:   Direct sequencing of exon 11 of *APC* using cycle sequencing with fluorescent dye terminators showing a G to A transition (arrowed) which removes the splice acceptor site prior to exon 14 (note that the sequencing has been carried out with the 3′ primer and as shown corresponds to the noncoding strand hence the mutation is shown as a C to T transition). Genomic DNA was sequenced, hence the presence of both normal and wild-type sequence. Courtesy of Alana Ward, DNA Laboratory, Regional Genetics Service, Birmingham Heartlands Hospital.

automated (*Figure 13.6*) making this a very rapid approach. However it is really only practical for the analysis of small genes or if there are only a small number of individuals to analyze.

13.2.4   Southern blotting

For over two decades until the development of PCR the basis of much of the analysis of genes and of gene tracking through families (see below) depended on the technique of Southern blotting, a technique named after its developer, Professor Ed Southern. All techniques subsequently developed for the analysis of RNA or protein have continued with this analogy for nomenclature hence, Northern (RNA), Western (protein) and even Southwestern (DNA/protein) blotting. Although the technique is no longer used as frequently, it still remains useful in some cases to identify large deletions or insertions, or for the analysis of some RFLPs and minisatellite markers.

The basis of Southern blotting is shown in *Figure 13.7*. DNA is extracted from the tissues or cells and digested into small fragments with restriction enzymes. These enzymes, which are isolated from bacteria, selectively cut DNA at sites dependent on specific nucleotide sequences in the genome (restriction sites; *Table 13.1*). Following digestion, these double-stranded DNA fragments are separated according to size by electrophoresis on an agarose gel. Treatment of the DNA in the gel with alkali produces single-stranded molecules, which are transferred or 'blotted' on to nitrocellulose or nylon membranes. The presence or absence of sequences of interest is then determined by incubating the membrane with the relevant DNA probe which has previously been labeled (either with an isotope such as ^{32}P, or with a molecule such as biotin or digoxigenin) and made single-stranded. Hybridization between the complementary sequences of the probe and target DNA is detected by autoradiography if ^{32}P has been used to label the probe or an enzyme–dye complex, in the case of biotinylated or digoxigenin labeled probes, allowing recognition of the DNA sequence under investigation and assessment of alterations in its size, copy number, etc. [1].

13.2.5   Gene tracking with RFLPs and microsatellites

Gene tracking is a method by which chromosomes can be tracked through a family to determine which one of a pair segregates with a disease gene. It depends on knowing on which chromosome and roughly where along its length a gene for a particular disease is located, and also that there is a polymorphic marker situated in the same region which can be used to 'tag' that chromosomal region. Traditionally this technique uses RFLP analysis but these days more usually involves the analysis of polymorphic microsatellite markers.

This technique has been used successfully for the diagnosis of many single gene disorders such as Duchenne muscular dystrophy or cystic fibrosis. In can-

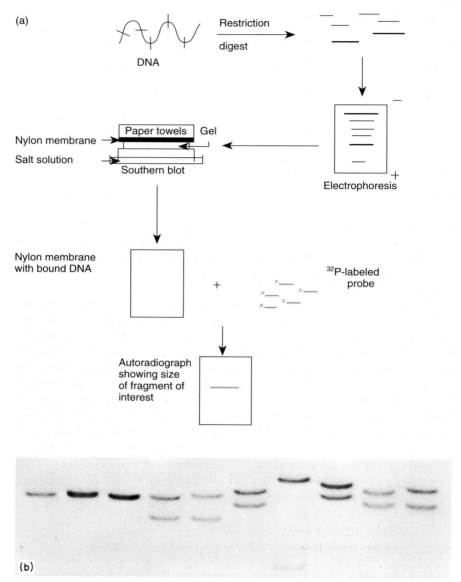

Figure 13.7:   (a) Southern blotting technique. (b) Example of the end-product of Southern blotting; an autoradiograph showing bands of different sizes.

cer research it has been used to track the defective gene in families with inherited cancers such as familial breast or colon cancer. The basis of the method is described first for RFLP analysis then compared with microsatellite analysis which is now used more frequently.

   In the example shown in *Figure 13.8,* if the DNA from individuals with dif-

Table 13.1:   Commonly used restriction enzymes

Name	Recognition sequence	Bacterial strain of origin
*Bam*HI	GGATCC CCTAGG	*Bacillus amyloliquefaciens*
*Eco*RI	GAATTC CTTAAG	*Escherichia coli*
*Hind*III	AAGCTT TTCGAA	*Hemophilus influenzae*
*Pst*I	CTGCAG GACGTC	*Providencia stuartii*
*Taq*I	TCGA AGCT	*Thermus aquaticus*

ferent DNA polymorphisms is digested with a restriction enzyme and then Southern blotted, individuals with chromosome A will have a 5 kb fragment detected by the probe, whereas individuals with chromosome B will have a 3 kb fragment detected by the same probe. The 5 kb fragment is arbitrarily called allele 1 and the 3 kb fragment allele 2.

For the detection of inherited diseases, the basis of the test is as follows: DNA from individuals in a family is digested with restriction enzymes and subjected to Southern blotting as described earlier. Each individual in the family is then given a genotype corresponding to the alleles present on either of the two chromosomes. *Figure 13.9* shows the result for three individuals using the probe and RFLP shown in *Figure 13.8* . Individual 1 is homozygous for allele 1, that is, has two copies of chromosome A. Individual 3 is also homozygous, but this time for allele 2. The individual shown in the center is a heterozygote, that is, has one copy of chromosome A and one of chromosome B. By entering these results on to the family pedigree along with results from any other members of the family available, it is possible to determine which allele, corre-

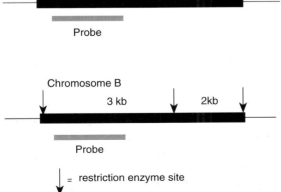

Figure 13.8:   Example of a RFLP showing one chromosome with a 5 kb DNA fragment and a second chromosome with a 3 kb fragment recognized by the same probe. Arrows indicate the restriction sites.

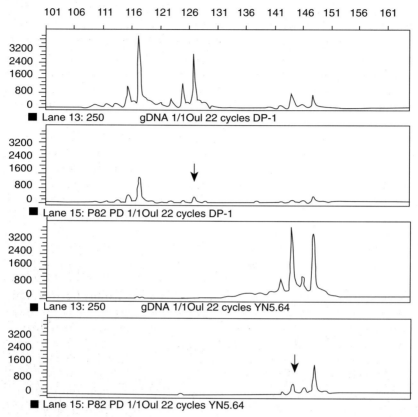

Figure 13.11: Microsatellite analysis using fluorescently labeled primers. The products were analyzed using an ABI 373 sequencer using GENESCAN software. The position on the x-axis corresponds to the size of the allele. The example shows the use of two microsatellite markers [LNS (DP-1): panels 1 and 2, YN5.64: panels 3 and 4] to study LOH around the *APC* gene. In the first and third panel, genomic DNA has been used as the template in the PCR reaction. In the second and fourth panel, target DNA was extracted from a colorectal tumor. The patient is heterozygous for the two markers in the genomic DNA but one allele is significantly reduced in the tumor DNA, indicating LOH. A small peak is still present in the tumor DNA because of the presence of normal cells in the tumor.

of the transcript to be determined and can also be used to detect abnormalities in the size of the mRNA [1].

13.2.7   In situ *hybridization*

This technique permits the direct analysis of sequences of DNA or RNA in tissues so that specific cells, populations of cells, or chromosomes can be exam-

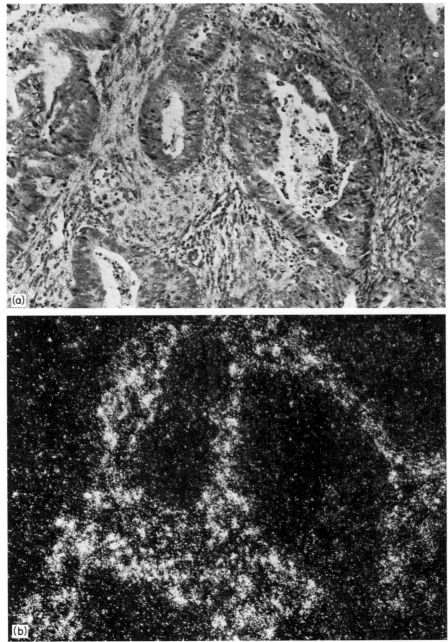

Figure 13.12: (a) Paraffin section of a colon cancer. Bright field illumination to show tissue morphology. (b) *In situ* hybridization using a ^{35}S labeled cDNA probe to detect type 1 collagen mRNA. Detection is by autoradiography. Dark field illumination shows silver grain distribution. Photographs courtesy of D.G. Powell, G.I. Carter and R.E. Hewitt, Department of Histopthology, University of Nottingham, Nottingham, UK.

ined [11]. The method involves the denaturation of DNA in a tissue section followed by the application of a labeled probe which is complementary to the sequence of interest. As with Southern blotting it is possible to use a radioactively labeled probe, but here the isotopes used are usually ^3H, ^{125}I or ^{35}S (*Figure 13.12*). In addition, as with Southern blotting techniques, it is now common to use probes labeled with biotin which can be detected by an enzymatic reaction. This technique has several advantages over radioisotopes, particularly speed, safety and the long-term stability of the labeled probe.

13.2.8   Comparative genomic hybridization

Comparative genomic hybridization is a method by which differences in copy number between tumor and normal DNA can be determined [12]. The two DNA samples are differentially labeled with two fluorochromes then hybridized to normal human chromosomes in a metaphase spread (*Figure 13.13*). The fluorescence signals are quantitated by image analysis and the ratio of one fluorochrome to the other determined along the length of the chromo-

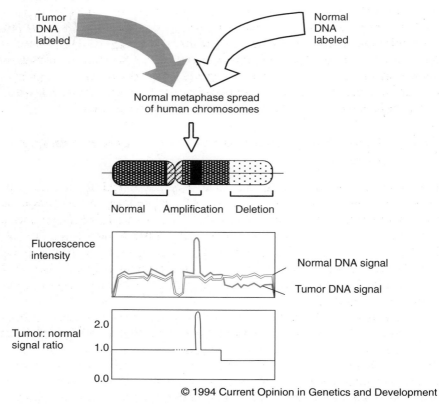

© 1994 Current Opinion in Genetics and Development

Figure 13.13:   A schematic representation of comparative genomic hybridization analysis. Reproduced from ref. 13 with permission from Current Biology Ltd.

some thereby allowing the determination of over or under representation of DNA segments in the tumor genome. Gains in DNA sequences are detected by increased signals from the fluorochrome labeling the tumor DNA whereas losses are detected by a signal from the normal DNA but not from the tumor DNA at the same site. This technique has been used to study copy number in both solid tumors and in leukemias. An adaptation of this approach, in which DNA from the tumor tissue is first amplified by degenerate oligonucleotide primed PCR (DOP-PCR), can also be used on formalin-fixed paraffin-embedded material and can give results even if only a few tumor cells are present.

13.3 Analysis of proteins

An alternative strategy used to examine the expression of genes is to look at the proteins themselves rather than at the DNA or RNA. This method has some advantages. In particular, it means that a retrospective analysis of tumor material can be made. This is made possible by the availability of paraffin blocks from surgically resected specimens. This material is not suitable for analysis of RNA although it is now possible to extract DNA from the blocks for analysis by PCR. The presence of oncoproteins in such tissue is most easily assessed by the application of antibodies. The most obvious advantage of this technique is the ability to test samples from individuals for whom the clinical outcome is known. In studies of the value of oncogenes as prognostic indicators this is invaluable as prospective studies could take years to perform. In order to study oncoprotein expression, however, antibodies have to be produced.

13.3.1 Production of antibodies to onco- or tumor suppressor proteins

In general, antibodies to the proteins have been raised against synthetic peptides rather than the native protein molecule. Short peptides, synthesized on the basis of known DNA sequences and coupled to protein carriers, have been used as immunogens. The region chosen for synthesis of the peptides is selected following the construction of a hydrophobicity plot and includes those residues thought to be exposed on the surface of the intact molecule. Both monoclonal and polyclonal antibodies have been produced in this way and the approach has found many useful applications [13].

A novel approach to producing antibodies specific for tumor cells has been to transfect the NIH-3T3 cell line with DNA fragments obtained from a human ALL and to use the transformants as immunogens to raise antibodies. Although the antigen has not been characterized in detail, at least one antibody produced in this way did not react with NIH-3T3 or with the majority of normal tissues tested but did react with certain tumors, including ALLs and sarcomas [14].

A list of some of the antibodies to oncoproteins currently available is shown in *Table 13.2*. The potential for the use of these antibodies is great, not only for examining the expression of the proteins in tissues but also for monitoring the levels of oncoproteins in body fluids and as potential carriers of therapeutic agents to cells, the so-called 'magic bullets'. Some of their applications have been discussed in previous chapters and the techniques used in their study are described in the following sections.

Table 13.2:   Oncogenes for which antibodies to oncoproteins are currently available

ABL	ERBB1
ERBB2	FMS
FOS	HRAS
KRAS	MOS
MYB	MYC
NRAS	SIS

This list is not comprehensive.

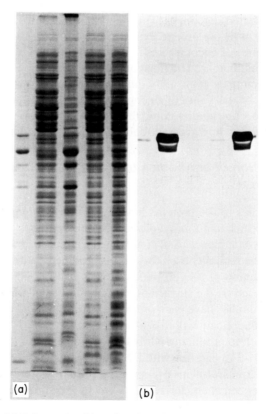

(a)   (b)

Figure 13.14:   (a) Polyacrylamide gel stained for total proteins. (b) Western blot of proteins extracted from tumor cells reacted with a monoclonal antibody to a tumor marker.

13.3.2 Western blotting

An overall examination of the size and quantity of oncoproteins in cells is possible by the technique of Western blotting. Proteins are extracted from cells and separated on polyacrylamide gels on the basis of size and/or charge. They are then transferred to nitrocellulose or nylon membranes. Unlike Southern blotting, transfer of proteins by passive diffusion is inefficient; Western blotting requires the application of an electric current for several hours. Following transfer, the protein of interest can be detected by incubation of the membrane

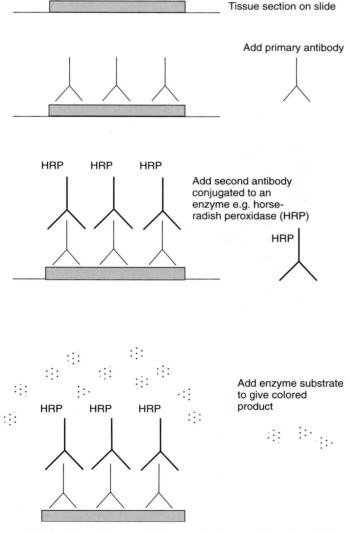

Tissue section on slide

Add primary antibody

Add second antibody conjugated to an enzyme e.g. horse-radish peroxidase (HRP)

Add enzyme substrate to give colored product

Figure 13.15 Flow diagram to illustrate the immunohistochemical technique.

in an antibody solution followed by detection with an enzymatically labeled second antibody [15] (*Figure 13.14*).

13.3.3 *Immunohistochemical techniques*

Immunohistochemical techniques are used widely throughout all areas of research to study expression of proteins in tissues, essentially by applying antibodies to thin sections of tissues, either frozen sections or sections of tissue which have been fixed and embedded in paraffin wax. Antibodies may be labeled directly with an enzyme such as horseradish peroxidase or alkaline phosphatase. Their binding sites can then be visualized by incubation of the tissue with a substrate which yields a colored product. More commonly, a second enzymatically labeled antibody specific for the primary antibody is used; this technique is described schematically in *Figure 13.15*. A typical result showing the distribution of the MYC oncoprotein in a section of gastric carcinoma is illustrated in *Figure 13.16*.

Some care has to be taken when interpreting the results of studies using immunohistochemistry, because the antigenic target for some antibodies

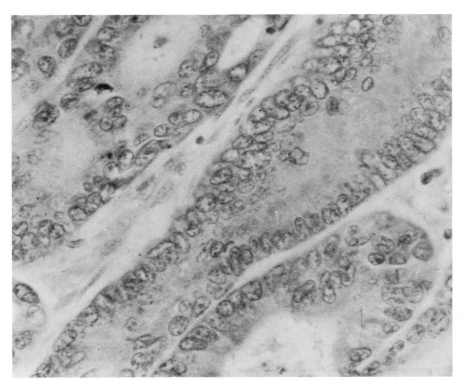

Figure 13.16:   A section of a gastric carcinoma reacted with an antibody to p62 (Myc1-6E10), showing cytoplasmic staining.

remains controversial. A number of antibodies have been produced to the *RAS* p21 protein. Some of these antibodies are reactive with both normal and mutant forms of the protein, for example, Y13 259, whereas others have been raised to a peptide containing the amino acid sequence corresponding to one of the forms of activated *RAS* and are specific for the mutated form of p21 [16]. A third antibody (RAP 5) has also been raised to a peptide reflecting the amino acid sequence of one of the activated forms of *RAS* but it cross-reacts with both the normal and mutated form of p21 [17]. Several antibodies (Myc1-6E10 and Myc1-9E10) to the MYC oncoprotein have been produced in a similar way, and again on further characterization it appears that one of these, Myc1-6E10, may not recognize the product of the *MYC* gene [18]. As mentioned in the chapters on breast, colorectal and gastrointestinal cancers, there has been some variation in results between centers using *p53* immunohistochemical staining. These are partly due to differences in technique and antibodies used and suggest that a degree of standardization is required. In addition, as mentioned

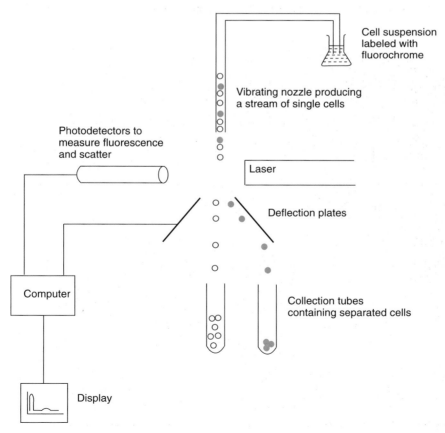

Figure 13.17:   Flow cytometry.

in Chapters 5, 6 and 7, comparisons between immunohistochemistry and detection of mutations at the DNA level have shown only a 69% correlation, leading some authors to doubt whether immunohistochemical techniques necessarily reflect *p53* mutations.

13.3.4 *Flow cytometry*

The development of the fluorescence-activated cell sorter (FACS) has provided an alternative method of studying oncogene expression; a method often referred to as flow cytometry. A wide range of particles can be analyzed using the FACS, including whole cells, nuclei, chromosomes and bacteria. Regardless of what is to be analyzed, the process is essentially the same. The particles are labeled with an appropriate fluorochrome and then passed individually through a laser beam. The emitted light for each individual particle is collected, measured and stored on computer. Other information such as particle volume and a measure of the granularity of particles can also be collected. In addition to the analysis, individual particles expressing particular pre-set parameters, such as fluorescence intensity, can be deflected electrostatically into collection tubes so that they can be analyzed further at the end of the experiment [19] (*Figure 13.17*). The fluorochrome used can be a fluorescent dye, such as fluorescein, coupled to an antibody specific for the protein under investigation, or it can be a nucleic-acid specific stain such as propidium iodide, ethidium bromide or DAPI. The whole procedure is extremely rapid, with

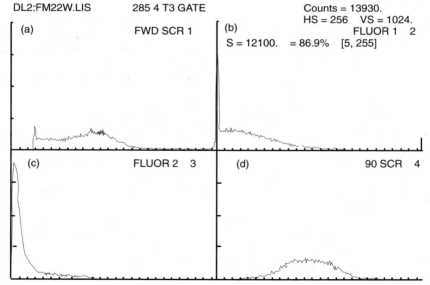

Figure 13.18:   Parameters measured by flow cytometer. (a) Distribution of cell sizes; (b) fluorescence with fluorochrome 1; (c) fluorescence with fluorochrome 2; and (d) granularity of cells (90° scatter).

thousands of particles being analyzed every second. A wide variety of parameters can subsequently be plotted either individually or against each other from the information obtained (*Figure 13.18*).

Nuclear oncogenes such as *MYC* and *FOS* have been extensively studied using flow cytometry [20]. Nuclei for analysis can be extracted from fresh tissues but can also be obtained from paraffin blocks. By dual labeling with a fluorochrome-labeled antibody plus a DNA-specific stain, it is possible to correlate the DNA content (ploidy status) with the expression of an oncogene in individual cells. The detection of aneuploidy in some tumors has been shown to correlate with a worse prognosis. The technique of combining information on oncogene expression with ploidy status may enhance its prognostic significance.

Flow cytometry is a very rapid method of determining the expression of oncoproteins on a cell-by-cell or nucleus-by-nucleus basis. It also has the advantage that, like immunocytochemical staining techniques, archival material can be used to obtain prognostic information. One drawback in the study of nuclei obtained from archival material, in addition to the problems previously mentioned concerning the specificity of the antibodies, is the apparent loss of reactivity of the nuclear antigens because of the harsh procedures necessary to extract nuclei [21]. This may make significant differences to the levels of oncoproteins detected in the tumors and makes quantitative analysis difficult on preserved tissue.

References

1. Sambrook, J., Fritsch, E.F. and Maniatis, T. (1989) *Molecular Cloning: a Laboratory Manual.* Cold Spring Harbor Laboratory Press, Cold Spring Harbor, NY.
2. Berger, S.L. and Kimmel, A.R. (1987) *Methods in Enzymology*, Vol. 152. *Guide to Molecular Cloning Techniques.* Academic Press, New York.
3. Ausubel, F.M. *et al.* (Eds) (1994) *Current Protocols in Molecular Biology.* John Wiley & Sons, New York.
4. Jeffreys, A.J. (1987) *Biochem. Soc. Trans.*, **15**, 309.
5. Weber, J.L. and May, P.E. (1989) *Am. J. Hum. Genet.*, **44**, 388.
6. Newton, C.R. and Graham, A. (1994) *PCR.* BIOS Scientific Publishers, Oxford.
7. Grompe, M. (1993) *Nature Genetics*, **5**, 111.
8. Bos, J.L., Fearon, E.R., Hamilton, S.R., Verlaan-de Vries, M., van Boom, J.H., van der Eb, A.J. and Vogelstein, B. (1987) *Nature*, **327**, 293.
9. Roest, P.A.M., Roberts, R.G., Sugino,S., van Ommen, G.J.B. and den Dunnen, J (1993) *Hum. Mol. Genet.*, **2**, 1719.
10. van der Luijt, R., Khan, P.M., Vasen, H. *et al.* (1994) *Genomics*, **20**, 1.
11. Buckle, V.J. and Craig, I. (1996) in *Human Genetic Diseases: a Practical Approach* (K.Davies, Ed.). IRL Press, Oxford, p. 85.
12. Buckle, V.J. and Kearney, L. (1994) *Curr. Opin. Genet. Dev.*, **4**, 374.
13. Tanaka, T., Slamon, D.J. and Cline, M.J. (1985) *Proc. Natl Acad. Sci. USA*, **82**, 3400.
14. Roth, J.A., Scuderi, P., Westin, E. and Gallo, R.C. (1985) *Surgery*, **96**, 264.
15. Burnette, W.N. (1981) *Analyt. Biochem.*, **112**, 195.

16.   Clark, R., Wong, G., Arnheim, N., Nitecki, D. and McCormick, F. (1985) *Proc. Natl Acad. Sci. USA*, **82**, 5280.
17.   Horan Hand, P., Thor, A., Wunderlich, D., Muraro, R., Caruso, A. and Schlom, J. (1984) *Proc. Natl Acad. Sci. USA*, **81**, 5227.
18.   Evan, G., Lewis, C.K., Ramsey, G. and Bishop, J.M. (1985) *Mol. Cell Biol.*, **5**, 3610.
19.   Young, B.D. (1986) in *Human Genetic Diseases: a Practical Approach* (K. Davies, Ed.). IRL Press, Oxford, p. 101.
20.   Stewart, C.C. (1989) *Arch. Pathol. Lab. Med.*, **113**, 634.
21.   Lincoln, S.T. and Bauer, K.D. (1989) *Cytometry*, **10**, 456.

Further reading

Strachan, T. and Read, A.P. (1996) *Human Molecular Genetics*. BIOS Scientific Publishers, Oxford.

Appendix A. Chromosomal location of the oncogenes

Oncogene	Chromosome	Oncogene	Chromosome
ABL	9q34	KRAS2	12p12.1
ABLL	1q24–q25	LCO	2q14
AKT1	14q32.3	LYN	8q13–qter
ARAF1	Xp11.4–p11.2	MAS1	6q24–q27
ARAF2	7p14–q21	MEL	19p13.2–cen
BCL2	18	MET	7q31–q32
CSF1R	5q33–q34	MOS	8q11 or 8q21–q23
EGFR	7p13–p12	MYB	6q22–q33
ELK1	Xp22.1–p11	MYC1	8q24
ELK2	14q32.3	MYCL1	1p32
ERBA2L	17q21–q22	MYCN	2p24
ERBAL2	19	NRAS1	1p22 and/or 1p13
ERBAL3	5	NRASL1	9p
ERBB2	17q11–q12	NRASL2	22
ERG	21q22.3	PDGFB	22q12.3–q13.1
ETS1	11q23.3	PIM1	6p21
ETS2	21q22.3	PVT1	8q24
FES	15q25–qter	RAF1	3p25
FGR	1p36.2–p36.1	RAF1P1	4p16.1
FOS	14q24.3	RALA	7p22–p15
GLI	12q13	REL	2p13–cen
HRAS	11p15.5	ROS1	6q21–q22
HRASP	Xpter–q26	RRAS	19
HRAS2P	Xp11.4–p11.2	SEA	11q13
HSTF1	11q13.3	SKI	1q21.1–q24
INT1	12q13	SPI1	11p12–p11.2
INT2	11q13	SRC	20q12–q13
INT4	17q21–q22	THRA1	17q11–q12
JUN	1p32–p31	THRB	3p24.1–p22
KIT	4p11–q22	YES1	18q21.3
KRAS1P	6p12–p11	YESP	22q11–q12

Data taken from *Cytogenetics and Cell Genetics*, Vol 51: *Human Gene Mapping 10* (1989), with permission from S. Karger AG, Basel.

Appendix B. Glossary

Alleles: different forms of a gene at the same position (locus) on a chromosome.

Aneuploid: any chromosome number which is a deviation from the exact multiple of the haploid number.

Antisense oligonucleotides: single-stranded oligonucleotides that bind in a complementary manner to mRNA, thereby modulating the transfer of information from gene to protein.

Apoptosis: a genetically encoded series of events that culminate in programed cell death or cellular suicide. The term is a morphological one which describes the fragmentation of a cell into membrane-bounded bodies.

Athymic mouse: *see* 'Nude' mouse.

Autocrine factor: a factor which is released by a cell which then stimulates the same cell (cf. Paracrine factor).

Autosome: any chromosome other than the sex chromosomes. In humans it refers to chromosomes 1–22.

cDNA: complementary DNA (DNA synthesized from mRNA using reverse transcriptase).

Centromere: *see* Chromosome.

Chimeric antibodies: antibodies assembled from diverse immunoglobulin gene segments that are not normally associated. Usually used to denote interspecies combinations where, for example, the variable region is of mouse origin and the constant region is of human origin.

Chromosome: a DNA-containing structure, visible under the microscope in dividing cells. At metaphase of mitosis, a chromosome comprises two identical sister chromatids joined at the centromere (*Figure A1*). In nondividing cells, chromosomes exist as extended single chromatids, not resolvable under the light microscope.

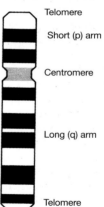

Figure A1: Chromosome structure.

Chromosome banding: chromosomes can be stained by a number of techniques. Routinely, they are stained either by G (Giemsa) or R (reverse) banding to give a specific banding pattern. By convention, the bands are numbered separately for the p and q arms, starting at the centromere and working towards the telomere.

Chromosome breakage syndromes: a group of disorders characterized by increased chromosome breakage in cultured cells. The group includes Bloom syndrome, xeroderma pigmentosa, ataxia telangiectasia and Fanconi anemia.

Chromosome walking: a method of moving from a characterized DNA clone known to be near a gene of interest in order to identify that gene. The technique utilizes the ability of the characterized clone to recognize, by homology, further clones containing part of the sequence of the initial clone, but which also contain adjacent DNA sequences along the chromosome.

Crossing over: the exchange of DNA between homologous chromosomes at meiosis I.

Cytokine: an extracellular signaling protein or peptide that acts as a local mediator in cell-to-cell communication.

Diploid: the chromosome number of somatic cells, which is twice that of the number found in gametes. In normal humans the diploid number is 46.

Dominant: any trait which is expressed in an heterozygote.

Enzyme-linked immunosorbent assay (ELISA): an assay dependent on the use of an enzyme-labeled antibody to measure antibody or antigen levels. The amount of labeled antibody bound to its target is usually determined by addition of the enzyme's substrate, which is degraded to give a colored product that can be measured spectrophotometrically.

Eukaryotic initiation signal: the codon at which translation begins in eukaryotes; usually AUG, which codes for methionine.

Exon: the region of a gene which contains the coding information.

Gene: a sequence of nucleotides in the DNA which encodes a single polypeptide. Contains both exons (coding sequences) and introns (noncoding sequences).

Gene therapy: the replacement of defective genes, or forms of corrective manipulation of genetic defects through genetic modification.

Genetic heterogeneity: phenotypically identical forms of a disease caused by mutations at two different loci.

Genotype: the genetic make-up of an individual. Used to describe alleles at a specific locus.

Germ-line mutation: a mutation present in the gametes which can be transmitted from parents to offspring.

Haploid: the chromosome number found in normal gametes.

Haplotype: a combination of the genotypes of several closely linked alleles which are inherited together.

Heat-shock protein: a protein produced by a cell in response to stress.

Hemizygote: an individual with one copy of an allele at one particular locus. Arises because only one chromosome (e.g. the X chromosome in males) or part of a chromosome (e.g. where there is a deletion) is present.

Heterozygote (heterozygous): an individual with two different alleles at the same locus on homologous chromosomes.

Homozygote (homozygous): identical alleles at the same locus on homologous chromosomes.

Hydrophobicity plot: a plot reflecting the distribution of hydrophilic or hydrophobic amino acid residues in a polypeptide. Used to predict those residues likely to be exposed on the surface of a molecule that may contribute to its antigenicity.

Immortal cell lines: those cells capable of continuous growth in tissue culture.

Immunocompromised mice: *see* 'Nude' mouse.

Imprinting: a mechanism in which the expression of a particular allele is dependent on parental origin (e.g. as seen in Beckwith–Wiedemann syndrome).

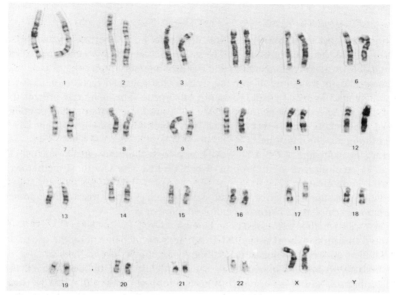

Figure A2: *Chromosomal make-up (karyotype) of a normal female. Chromosomes are laid out according to size and banding pattern from chromosomes 1 to 22 plus two X chromosomes.*

Interstitial deletion: deletion of part of a chromosome between two breakpoints.

Intron: a noncoding sequence, found in the eukaryotic genome, which is spliced out from the primary RNA transcript leading to the production of mature mRNA.

Karyotype: the chromosome make-up of an individual (*Figure A2*).

Leucine zipper: a helical stretch of amino acids rich in leucine residues, occuring at the turn of the helix. Two such helices can dimerize to form a coiled coil.

Linkage: the inheritance of two or more genes as a single unit because of their close proximity on the chromosome and not because of chance.

Li–Fraumeni syndrome: a dominantly inherited condition that carries with it a markedly increased risk of developing many malignancies, in particular soft tissue sarcomas, brain tumors, osteosarcomas, leukemias, adrenocortical carcinomas and carcinomas of the breast.

Liposomes: microscopic artificial phospholipid vesicles.

Locus: the chromosomal location defining the position of a gene.

Loss of heterozygosity: the loss of one allele of a marker in a tissue, usually tumor, when the individual is heterozygous in constitutional cells.

Meiosis: reduction cell division, resulting in the production of gametes each with a single copy of each chromosome.

Mendelian inheritance: the process by which individuals inherit one of each pair of chromosomes from each parent.

Messenger RNA (mRNA): the product of gene transcription which is subsequently translated into protein.

Microsatellite instability: an increase in the number of alleles for a microsatellite marker in tumor tissue compared with constitutional cells.

Mitosis: somatic cell division. The product is two daughter progeny, each with identical copies of every chromosome.

Monoclonal antibody (hybridoma): an antibody produced by a hybrid cell which is made by fusing a stimulated B cell with a myeloma cell line. The fused cell is clonal and an antibody with a single isotope and single specificity is produced. Because the myeloma cell is immortal, it can be grown continuously in culture, providing a continuous supply of identical antibodies.

Mutation: a heritable alteration in the DNA sequence.

'Nude' mouse: a mouse which is immunologically deficient. It can therefore be used to grow tumor cells from humans and other animals without rejection.

Oligonucleotide: a short sequence of nucleotides. May be synthesized chemically.

Paracrine factor: a factor released by one cell which affects others in the immediate vicinity.

Penetrance: the frequency of phenotypic expression of a gene mutation in affected individuals.

Peptide: a short amino acid sequence.

Philadelphia chromosome: a small chromosome produced by a balanced 9;22 translocation, particularly characteristic of chronic myeloid leukemia.

Point mutation: a mutation causing a single base change in the DNA sequence.

Polyclonal antiserum: an antibody 'cocktail' comprising a range of antibody isotypes recognizing a number of antigenic determinants on a target molecule. Usually isolated from the blood of an immunized animal. The source of the antiserum has a finite lifespan unlike the source of monoclonal antibodies.

Polymorphism: two or more alleles in a population, each of which is present at a frequency greater than that expected by mutational events.

Positional cloning: the cloning of a gene once it has been localized to a specific chromosomal region.

Promoter: sequences 5′ of a gene which regulate the initiation of transcription.

Radioimmunoassay: assay which depends on the use of a radioisotopically labeled antibody to measure either antibody or antigen levels.

Recessive: any trait which is expressed only in homozygotes.

Recombinant (immuno)toxins: single polypeptide chains with a recognition domain (e.g. peptide hormone, growth factor. cytokine or single chain antibody) which are genetically fused to a truncated toxin gene. For example, the gene for a toxin is spliced into a plasmid vector next to an antibody gene, resulting in a fused recombinant immunotoxin protein product.

Recombination: crossing over between loci on homologous chromosomes resulting in a new combination of linked genes.

Reverse genetics: the process by which a gene is mapped and eventually cloned without prior knowledge of the biochemical nature of the gene product.

Reverse transcriptase: an enzyme which produces a DNA sequence from an RNA template.

Sex chromosomes: X and Y chromosomes.

Somatic cell: any cell in the body except the gametes.

Sporadic cancer: a nonheritable form of cancer.

Telomere: *see* Chromosome.

Transcription: the synthesis of RNA from DNA using RNA polymerase.

Transfection: a method of transfering a DNA sequence into a cell.

Transgenic mouse: a mouse which has had foreign DNA inserted into the germ-line.

Transition: the most common type of point mutation in which one pyrimidine is substituted for the other, or in which one purine is substituted for the other (i.e. G–C exchanged for an A–T pair or vice versa).

Translation: the synthesis of a protein from an mRNA template.

Translocation: the transfer of chromosomal regions between nonhomologous chromosomes. May be balanced or unbalanced. Usually abbreviated to t followed by the chromosome numbers involved [e.g. t(8;14) which describes the translocation seen in Burkitt's lymphoma].

Transversion: the less common type of point mutation in which a purine is replaced by a pyrimidine or vice versa (e.g. A–T becomes a T–A or a C–G pair)

Tumor-associated antigen: an antigen found primarily associated with tumors but also expressed on normal and fetal tissues.

Xenograft: a graft across a species barrier (e.g. a human tumor growing in an athymic nude mouse).

Zinc finger protein: a transcription-regulating protein which has finger-like structures containing a zinc atom.

Index